Modern philosophy of science has paid great attention to an understanding of scientific practice, in contrast to the earlier concentration on 'scientific method'. The work of Karl Popper, Thomas Kuhn and Imre Lakatos has provided varying accounts of what this practice is. Paul Feyerabend goes beyond this position: he argues that the most successful scientific inquiries have never proceeded according to the rational method at all. He examines in detail the arguments which Galileo used to defend the Copernican revolution in physics, and shows that his success depended not on rational argument but on a mixture of subterfuge, rhetoric and propaganda.

Claiming that anarchism must now replace rationalism in the theory of knowledge, Feyerabend argues that intellectual progress can only be achieved by stressing the creativity and wishes of the scientist rather than the method and authority of science. In the latter half of the book he examines Popper's 'critical rationalism' and the attempt by Lakatos to construct a methodology which allows the scientist his freedom without threatening scientific 'law and order'. Rejecting both attempts to shore up rationalism, he looks forward to the 'withering away of reason' and maintains that 'the only principle which does not inhibit progress is anything goes'.

Paul Feyerabend

Verso

Against Method

Outline of an anarchistic theory of knowledge*

* For some comments on the term 'anarchism' as used cf. footnote 12 of the *Introduction*, and chapter 16, text to footnotes 16ff.

First published, 1975

Verso Edition first published, 1978

Second impression, 1979
Third impression, 1980
Fourth impression, 1982
Verso, 15 Greek Street, London W1

Printed in Great Britain by
Redwood Burn Limited Trowbridge, Wiltshire
and bound by Pegasus Bookbinding, Melksham, Wiltshire

SBN 86091 700 2 (paper)
SBN 902308 91 2 (cloth)

To

IMRE LAKATOS

Friend and fellow-anarchist

This essay is the first part of a book on rationalism that was to be written by Imre Lakatos and myself. I was to attack the rationalist position, Imre was to restate and to defend it, making mincemeat of me in the process. Taken together, the two parts were supposed to give an account of our long debate concerning these matters that had started in 1964, had continued, in letters, lectures, telephone calls, papers, almost to the last day of Imre's life and had become a natural part of my daily routine. The origin explains the style of the essay: it is a long and rather personal *letter* to Imre and every wicked phrase it contains was written in anticipation of an even more wicked reply from the recipient. It is also clear that as it stands the book is sadly incomplete. It lacks the most important part: the reply of the person to whom it is addressed. I still publish it as a testimony to the strong and exhilarating influence Imre Lakatos has had on all of us.

The paperback edition is an unchanged reprint of the hardcover edition. Replies to criticism and further developments will be contained in a separate volume to appear late in 1978. Some of the additional material has already been published in the German edition *Wider den Methodenzwang*, Frankfurt 1976. The reader who wants to know right away what I think about the many people who have cursed my book in print should consult my *Science in a Free Society* (NLB, 1978).

Paul K. Feyerabend

Analytical Index

Analytical Index

Being a Sketch of the Main Argument

Items:

Introduction

Science is an essentially anarchistic enterprise: theoretical anarchism is more humanitarian and more likely to encourage progress than its law-and-order alternatives.

1 page 23

This is shown both by an examination of historical episodes and by an abstract analysis of the relation between idea and action. The only principle that does not inhibit progress is: *anything goes.*

2 page 29

For example, we may use hypotheses that contradict well-confirmed theories and/or well-established experimental results. We may advance science by proceeding counterinductively.

3 page 35

The consistency condition which demands that new hypotheses agree with accepted *theories* is unreasonable because it preserves the older theory, and not the better theory. Hypotheses contradicting well-confirmed theories give us evidence that cannot be obtained in any other way. Proliferation of theories is beneficial for science, while uniformity impairs its critical power. Uniformity also endangers the free development of the individual.

4 page 47

There is no idea, however ancient and absurd, that is not capable of improving our knowledge. The whole history of thought is absorbed into science and is used for improving every single theory. Nor is political interference rejected. It may be needed to overcome the chauvinism of science that resists alternatives to the status quo.

5 page 55

No theory ever agrees with all the *facts* in its domain, yet it is not always the theory that is to blame. Facts are constituted by older ideologies, and a clash between facts and theories may be proof of progress. It is also a first step in our attempt to find the principles implicit in familiar observational notions.

6 page 69

As an example of such an attempt I examine the *tower argument* which the Aristotelians used to refute the motion of the earth. The argument involves *natural interpretations* – ideas so closely connected with observations that it needs a special effort to realize their existence and to determine their content. Galileo identifies the natural interpretations which are inconsistent with Copernicus and replaces them by others.

7 page 81

The new natural interpretations constitute a new and highly abstract observation language. They are introduced *and concealed* so that one fails to notice the change that has taken place (method of anamnesis). They contain the idea of the *relativity of all motion* and the *law of circular inertia*.

8 page 93

Initial difficulties caused by the change are defused by *ad hoc hypotheses*, which thus turn out occasionally to have a positive function; they give new theories a breathing space, and they indicate the direction of future research.

9 page 99

In addition to natural interpretations, Galileo also changes *sensations* that seem to endanger Copernicus. He admits that there are such sensations, he praises Copernicus for having disregarded them, he claims to have removed them with the help of the *telescope*. However, he offers no *theoretical* reasons why the telescope should be expected to give a true picture of the sky.

appendix 1 page 109

appendix 2 page 112

10 page 121

Nor does the initial *experience* with the telescope provide such reasons. The first telescopic observations of the sky are indistinct, indeterminate, contradictory and in conflict with what everyone can see with his unaided eyes. And, the only theory that could have helped to separate telescopic illusions from veridical phenomena was refuted by simple tests.

11 page 141

On the other hand, there are some telescopic phenomena which are plainly Copernican. Galileo introduces these phenomena as independent evidence for Copernicus while the situation is rather that one refuted view – Copernicanism – has a certain similarity with phenomena emerging from another refuted view – the idea that telescopic phenomena are faithful images of the sky. Galileo prevails because of his style and his clever techniques of persuasion, because he writes in Italian rather than in Latin, and because he appeals to people who are temperamentally opposed to the old ideas and the standards of learning connected with them.

12 page 145

Such 'irrational' methods of support are needed because of the 'uneven development' (Marx, Lenin) of different parts of science. Copernicanism and other essential ingredients of modern science survived only because reason was frequently overruled in their past.

13 page 163

Galileo's method works in other fields as well. For example, it can be used to eliminate the existing arguments against materialism, and to put an end to the *philosophical* mind/body problem (the corresponding *scientific* problems remain untouched, however).

14 page 165

The results obtained so far suggest abolishing the distinction between a context of discovery and a context of justification and disregarding the related distinction between observational terms and theoretical terms. Neither distinction plays a role in scientific practice. Attempts to enforce them would have disastrous consequences.

15 page 171

Finally, the discussion in Chapters 6–13 shows that Popper's version of Mill's pluralism is not in agreement with scientific practice and would destroy science as we know it. Given science, reason cannot be universal and unreason cannot be excluded. This feature of science calls for an anarchistic epistemology. The realization that science is not sacrosanct, and that the debate between science and myth has ceased without having been won by either side, further strengthens the case for anarchism.

16 page 181

Even the ingenious attempt of Lakatos to construct a methodology that (a) does not issue orders and yet (b) puts restrictions upon our knowledge-increasing activities, does not escape this conclusion. For Lakatos' philosophy appears liberal only because it is an *anarchism in disguise*. And his standards which are abstracted from modern science cannot be regarded as neutral arbiters in the issue between modern science and Aristotelian science, myth, magic, religion, etc.

appendix 3 page 215

appendix 4 page 221

17 page 223

Moreover, these standards, which involve a comparison of content classes, are not always *applicable*. The content classes of certain theories are incomparable in the sense that none of the usual logical relations (inclusion, exclusion, overlap) can be said to hold between them. This occurs when we compare myths with science. It also occurs in the most advanced, most general and therefore most mythological parts of science itself.

appendix 5 page 286

18 page 295

Thus science is much closer to myth than a scientific philosophy is prepared to admit. It is one of the many forms of thought that have been developed by man, and not necessarily the best. It is conspicuous, noisy, and impudent, but it is inherently superior only for those who have already decided in favour of a certain ideology, or who have accepted it without having ever examined its advantages and its limits. And as the accepting and rejecting of ideologies should be left to the individual it follows that the separation of state and *church* must be supplemented by the separation of state and *science*, that most recent, most aggressive, and most dogmatic religious institution. Such a separation may be our only chance to achieve a humanity we are capable of, but have never fully realized.

indices page 311

Introduction

Ordnung ist heutzutage meistens dort,
wo nichts ist.
Es ist eine Mangelerscheinung.

BRECHT

Science is an essentially anarchistic enterprise: theoretical anarchism is more humanitarian and more likely to encourage progress than its law-and-order alternatives.

The following essay is written in the conviction that *anarchism*, while perhaps not the most attractive *political* philosophy, is certainly excellent medicine for *epistemology*, and for the *philosophy of science*.

The reason is not difficult to find.

'History generally, and the history of revolutions in particular, is always richer in content, more varied, more many-sided, more lively and subtle than even' the best historian and the best methodologist can imagine.[1] History is full of 'accidents and conjunctures and curious juxtapositions of events'[2] and it demonstrates to us the 'complexity of human change and the unpredictable character of the ultimate consequences of any given act or decision of men'.[3] Are we really to believe that the naive and simple-minded rules which methodologists take as

1. 'History as a whole, and the history of revolutions in particular, is always richer in content, more varied, more multiform, more lively and ingenious than is imagined by even the best parties, the most conscious vanguards of the most advanced classes' (V. I. Lenin, 'Left-Wing Communism – An Infantile Disorder', *Selected Works*, Vol. 3, London, 1967, p. 401). Lenin is addressing parties and revolutionary vanguards rather than scientists and methodologists; the lesson, however, is the same. cf. footnote 5.

2. Herbert Butterfield, *The Whig Interpretation of History*, New York, 1965, p. 66.

3. ibid., p. 21.

their guide are capable of accounting for such a 'maze of interactions'?[4] And is it not clear that successful *participation* in a process of this kind is possible only for a ruthless opportunist who is not tied to any particular philosophy and who adopts whatever procedure seems to fit the occasion?

This is indeed the conclusion that has been drawn by intelligent and thoughtful observers. 'Two very important practical conclusions follow from this [character of the historical process],' writes Lenin,[5] continuing the passage from which I have just quoted. 'First, that in order to fulfil its task, the revolutionary class [i.e. the class of those who want to change either a part of society such as science, or society as a whole] must be able to master *all* forms or aspects of social activity without exception [it must be able to understand, and to apply, not only one particular methodology, but any methodology, and any variation thereof it can imagine] . . .; second [it] must be ready to pass from one to another in the quickest and most unexpected manner.' 'The external conditions,' writes Einstein,[6] 'which are set for [the scientist] by the facts of experience do not permit him to let himself be too much restricted, in the construction of his conceptual world, by the adherence to an epistemological system. He therefore, must appear to the systematic epistemologist as a type of unscrupulous opportunist. . . .' A complex medium containing surprising and unforeseen developments demands complex procedures and defies analysis on the basis of rules which have been set up in advance and without regard to the ever-changing conditions of history.

4. ibid., p. 25, cf. Hegel, *Philosophie der Geschichte*, *Werke*, Vol. 9, ed. Edward Gans, Berlin, 1837, p. 9: 'But what experience and history teach us is this, that nations and governments have never learned anything from history, or acted according to rules that might have derived from it. Every period has such peculiar circumstances, is in such an individual state, that decisions will have to be made, and decisions *can* only be made, in it and out of it.' – 'Very clever'; 'shrewd and very clever'; 'NB' writes Lenin in his marginal notes to this passage. (*Collected Works*, Vol. 38, London, 1961, p. 307.)

5. ibid. We see here very clearly how a few substitutions can turn a political lesson into a lesson for *methodology*. This is not at all surprising. Methodology and politics are both means for moving from one historical stage to another. The only difference is that the standard methodologies disregard the fact that history constantly produces new features. We also see how an individual, such as Lenin, who is not intimidated by traditional boundaries and whose thought is not tied to the ideology of a profession, can give useful advice to everyone, philosophers of science included.

6. Albert Einstein, *Albert Einstein: Philosopher Scientist*, ed. P. A. Schilpp, New York, 1951, pp. 683f.

Now it is, of course, possible to simplify the medium in which a scientist works by simplifying its main actors. The history of science, after all, does not just consist of facts and conclusions drawn from facts. It also contains ideas, interpretations of facts, problems created by conflicting interpretations, mistakes, and so on. On closer analysis we even find that science knows no 'bare facts' at all but that the 'facts' that enter our knowledge are already viewed in a certain way and are, therefore, essentially ideational. This being the case, the history of science will be as complex, chaotic, full of mistakes, and entertaining as the ideas it contains, and these ideas in turn will be as complex, chaotic, full of mistakes, and entertaining as are the minds of those who invented them. Conversely, a little brainwashing will go a long way in making the history of science duller, simpler, more uniform, more 'objective' and more easily accessible to treatment by strict and unchangeable rules.

Scientific education as we know it today has precisely this aim. It simplifies 'science' by simplifying its participants: first, a domain of research is defined. The domain is separated from the rest of history (physics, for example, is separated from metaphysics and from theology) and given a 'logic' of its own. A thorough training in such a 'logic' then conditions those working in the domain; it makes *their actions* more uniform and it freezes large parts of the *historical process* as well. Stable 'facts' arise and persevere despite the vicissitudes of history. An essential part of the training that makes such facts appear consists in the attempt to inhibit intuitions that might lead to a blurring of boundaries. A person's religion, for example, or his metaphysics, or his sense of humour (his *natural* sense of humour and not the inbred and always rather nasty kind of jocularity one finds in specialized professions) must not have the slightest connection with his scientific activity. His imagination is restrained, and even his language ceases to be his own.[7] This is again reflected in the nature of scientific 'facts' which are experienced as being independent of opinion, belief, and cultural background.

It is thus *possible* to create a tradition that is held together by strict rules, and that is also successful to some extent. But is it *desirable* to support such a tradition to the exclusion of everything else? Should we

7. For the deterioration of language that follows any increase of professionalism cf. my essay 'Experts in a Free Society', *The Critic*, November/December 1970.

transfer to it the sole rights for dealing in knowledge, so that any result that has been obtained by other methods is at once ruled out of court? This is the question I intend to ask in the present essay. And to this question my answer will be a firm and resounding NO.

There are two reasons why such an answer seems to be appropriate. The first reason is that the world which we want to explore is a largely unknown entity. We must, therefore, keep our options open and we must not restrict ourselves in advance. Epistemological prescriptions may look splendid when compared with other epistemological prescriptions, or with general principles – but who can guarantee that they are the best way to discover, not just a few isolated 'facts', but also some deep-lying secrets of nature? The second reason is that a scientific education as described above (and as practised in our schools) cannot be reconciled with a humanitarian attitude. It is in conflict 'with the cultivation of individuality which alone produces, or can produce, well-developed human beings'[8]; it 'maims by compression, like a Chinese lady's foot, every part of human nature which stands out prominently, and tends to make a person markedly different in outline'[9] from the ideals of rationality that happen to be fashionable in science, or in the philosophy of science. The attempt to increase liberty, to lead a full and rewarding life, and the corresponding attempt to discover the secrets of nature and of man entails, therefore, the rejection of all universal standards and of all rigid traditions. (Naturally, it also entails the rejection of a large part of contemporary science.)

It is surprising to see how rarely the stultifying effect of 'the Laws of Reason' or of scientific practice is examined by professional anarchists. Professional anarchists oppose any kind of restriction and they demand that the individual be permitted to develop freely, unhampered by laws, duties or obligations. And yet they swallow without protest all the severe standards which scientists and logicians impose upon research and upon any kind of knowledge-creating and knowledge-changing activity. Occasionally, the laws of scientific method, or what are thought to be the laws of scientific method by a particular writer, are even integrated into

8. John Stuart Mill, 'On Liberty', *The Philosophy of John Stuart Mill*, ed. Marshall Cohen, New York, 1961, p. 258.

9. ibid., p. 265.

anarchism itself. 'Anarchism is a world concept based upon a mechanical explanation of all phenomena,' writes Kropotkin.[10] 'Its method of investigation is that of the exact natural sciences . . . the method of induction and deduction.' 'It is not so clear,' writes a modern 'radical' professor at Columbia,[11] 'that scientific research demands an absolute freedom of speech and debate. Rather the evidence suggests that certain kinds of unfreedom place no obstacle in the way of science. . . .'

There are certainly some people to whom this is 'not so clear'. Let us, therefore, start with our outline of an anarchistic methodology and a corresponding anarchistic science.[12] There is no need to fear that the diminished concern for law and order in science and society that characterizes an anarchism of this kind will lead to chaos. The human nervous

10. Peter Alexeivich Kropotkin, 'Modern Science and Anarchism', *Kropotkin's Revolutionary Pamphlets*, ed. R. W. Baldwin, New York, 1970, pp. 150–2. 'It is one of Ibsen's great distinctions that nothing was valid for him but science.' B. Shaw, *Back to Methuselah*, New York, 1921, xcvii. Commenting on these and similar phenomena Strindberg writes (*Antibarbarus*): 'A generation that had the courage to get rid of God, to crush the state and church, and to overthrow society and morality, still bowed before Science. And in Science, where freedom ought to reign, the order of the day was "believe in the authorities or off with your head".'

11. R. P. Wolff, *The Poverty of Liberalism*, Boston, 1968, p. 15. For a more detailed criticism of Wolff see footnote 52 of my essay 'Against Method' in *Minnesota Studies in the Philosophy of Science*, Vol. 4, Minneapolis, 1970.

12. When choosing the term 'anarchism' for my enterprise I simply followed general usage. However anarchism, as it has been practised in the past and as it is being practised today by an ever increasing number of people has features I am not prepared to support. It cares little for human lives and human happiness (except for the lives and the happiness of those who belong to some special group); and it contains precisely the kind of Puritanical dedication and seriousness which I detest. (There are some exquisite exceptions such as Cohn-Bendit, but they are in the minority.) It is for these reasons that I now prefer to use the term *Dadaism*. A Dadaist would not hurt a fly – let alone a human being. A Dadaist is utterly unimpressed by any serious enterprise and he smells a rat whenever people stop smiling and assume that attitude and those facial expressions which indicate that something important is about to be said. A Dadaist is convinced that a worthwhile life will arise only when we start taking things *lightly* and when we remove from our speech the profound but already putrid meanings it has accumulated over the centuries ('search for truth'; 'defence of justice'; 'passionate concern'; etc., etc.) A Dadaist is prepared to initiate joyful experiments even in those domains where change and experimentation seem to be out of the question (example: the basic functions of language). I hope that having read the pamphlet the reader will remember me as a flippant Dadaist and *not* as a serious anarchist. cf. footnote 4 of chapter 2.

system is too well organized for that.[13] There may, of course, come a time when it will be necessary to give reason a temporary advantage and when it will be wise to defend its rules to the exclusion of everything else. I do not think that we are living in such a time today.

13. Even in undetermined and ambiguous situations, uniformity of action is soon achieved and adhered to tenaciously. See Muzafer Sherif, *The Psychology of Social Norms*, New York, 1964.

1

This is shown both by an examination of historical episodes and by an abstract analysis of the relation between idea and action. The only principle that does not inhibit progress is: anything goes.

The idea of a method that contains firm, unchanging, and absolutely binding principles for conducting the business of science meets considerable difficulty when confronted with the results of historical research. We find then, that there is not a single rule, however plausible, and however firmly grounded in epistemology, that is not violated at some time or other. It becomes evident that such violations are not accidental events, they are not results of insufficient knowledge or of inattention which might have been avoided. On the contrary, we see that they are necessary for progress. Indeed, one of the most striking features of recent discussions in the history and philosophy of science is the realization that events and developments, such as the invention of atomism in antiquity, the Copernican Revolution, the rise of modern atomism (kinetic theory; dispersion theory; stereochemistry; quantum theory), the gradual emergence of the wave theory of light, occurred only because some thinkers either *decided* not to be bound by certain 'obvious' methodological rules, or because they *unwittingly broke* them.

This liberal practice, I repeat, is not just a *fact* of the history of science. It is both reasonable and *absolutely necessary* for the growth of knowledge. More specifically, one can show the following: given any rule, however 'fundamental' or 'necessary' for science, there are always circumstances when it is advisable not only to ignore the rule, but to adopt its opposite. For example, there are circumstances when it is advisable to introduce, elaborate, and defend *ad hoc* hypotheses, or hypotheses which contradict well-established and generally accepted experimental results, or hypotheses whose content is smaller than the content of the existing and empirically adequate alternative, or self-inconsistent hypotheses, and so on.[1]

There are even circumstances – and they occur rather frequently – when *argument* loses its forward-looking aspect and becomes a hindrance to progress. Nobody would claim that the teaching of *small children* is exclusively a matter of argument (though argument may enter into it, and should enter into it to a larger extent than is customary), and almost everyone now agrees that what looks like a result of reason – the mastery of a language, the existence of a richly articulated perceptual world, logical ability – is due partly to indoctrination and partly to a process of *growth* that proceeds with the force of natural law. And where arguments *do* seem to have an effect, this is more often due to their *physical repetition* than to their *semantic content*.

Having admitted this much, we must also concede the possibility of non-argumentative growth in the *adult* as well as in (the theoretical parts of) *institutions* such as science, religion, prostitution, and so on. We certainly cannot take it for granted that what is possible for a small child – to acquire new modes of behaviour on the slightest provocation, to slide into them without any noticeable effort – is beyond the reach of his elders. One should rather expect that catastrophic changes in the physical environment, wars, the breakdown of encompassing systems of morality, political revolutions, will transform adult reaction patterns as well, including important patterns of argumentation. Such a transformation may again be an entirely natural process and the only function of a rational argument may lie in the fact that it increases the mental tension that precedes *and causes* the behavioural outburst.

1. One of the few thinkers to understand this feature of the development of knowledge was Niels Bohr: '. . . he would never try to outline any finished picture, but would patiently go through all the phases of the development of a problem, starting from some apparent paradox, and gradually leading to its elucidation. In fact, he never regarded achieved results in any other light than as starting points for further exploration. In speculating about the prospects of some line of investigation, he would dismiss the usual consideration of simplicity, elegance or even consistency with the remark that such qualities can only be properly judged *after* [my italics] the event. . . .' L. Rosenfeld in *Niels Bohr. His Life and Work as seen by his Friends and Colleagues*, ed. S. Rosental, New York, 1967, p. 117. Now science is never a completed process, therefore it is always 'before' the event. Hence simplicity, elegance or consistency are *never* necessary conditions of (scientific) practice.

Now, if there are events, not necessarily arguments which *cause* us to adopt new standards, including new and more complex forms of argumentation, is it then not up to the defenders of the *status quo* to provide, not just counter-arguments, but also contrary *causes*? ('Virtue without terror is ineffective,' says Robespierre.) And if the old forms of argumentation turn out to be too weak a cause, must not these defenders either give up or resort to stronger and more 'irrational' means? (It is very difficult, and perhaps entirely impossible, to combat the effects of brainwashing by argument.) Even the most puritanical rationalist will then be forced to stop reasoning and to use *propaganda* and *coercion*, not because some of his *reasons* have ceased to be valid, but because the *psychological conditions* which make them effective, and capable of influencing others, have disappeared. And what is the use of an argument that leaves people unmoved?

Of course, the problem never arises quite in this form. The teaching of standards and their defence never consists merely in putting them before the mind of the student and making them as *clear* as possible. The standards are supposed to have maximal *causal efficacy* as well. This makes it very difficult indeed to distinguish between the *logical force* and the *material effect* of an argument. Just as a well-trained pet will obey his master no matter how great the confusion in which he finds himself, and no matter how urgent the need to adopt new patterns of behaviour, so in the very same way a well-trained rationalist will obey the mental image of *his* master, he will conform to the standards of argumentation he has learned, he will adhere to these standards no matter how great the confusion in which he finds himself, and he will be quite incapable of realizing that what he regards as the 'voice of reason' is but a *causal after-effect* of the training he has received. He will be quite unable to discover that the appeal to reason to which he succumbs so readily is nothing but a *political manoeuvre*.

That interests, forces, propaganda and brainwashing techniques play a much greater role than is commonly believed in the growth of our knowledge and in the growth of science, can also be seen from an analysis of the *relation between idea and action*. It is often taken for granted that a clear and distinct understanding of new ideas precedes, and should precede, their formulation and their institutional expression. (An investigation starts with a problem, says Popper.) *First*, we have an idea, or a problem, *then* we act, i.e. either speak, or build, or destroy. Yet this is certainly not the way in which small children develop. They use words, they combine them, they play with them, until they grasp a meaning that has so far been beyond their reach. And the initial playful activity is an essential prerequisite of the final act of understanding. There is no reason why this mechanism should cease to function in the adult. We must expect, for example, that the *idea* of liberty could be made clear only by means of the very same actions, which were supposed to *create* liberty. Creation of a *thing*, and creation plus full understanding of a *correct idea* of the thing, *are very often parts of one and the same indivisible process* and cannot be separated without bringing the process to a stop. The process itself is not guided by a well-defined programme, and cannot be guided by such a programme, for it contains the conditions for the realization of all possible programmes. It is guided rather by a vague urge, by a 'passion' (Kierkegaard). The passion gives rise to specific behaviour which in turn creates the circumstances and the ideas necessary for analysing and explaining the process, for making it 'rational'.

The development of the Copernican point of view from Galileo to the 20th century is a perfect example of the situation I want to describe. We start with a strong belief that runs counter to contemporary reason and contemporary experience. The belief spreads and finds support in other beliefs which are equally unreasonable, if not more so (law of inertia; the telescope). Research now gets deflected in new directions, new kinds of instruments are built, 'evidence' is related to theories in new ways until there arises an ideology that is rich enough to provide independent arguments for any particular part of it and mobile enough to find such arguments whenever they seem to be required. We can say today that Galileo was on the right track, for his persistent pursuit of what once seemed to be a silly cosmology has by now created the material needed to defend it against all those who will accept a view only if it is told in a certain way and who will trust it only if it contains certain magical phrases, called 'observational reports'. And this is not an exception – it is the normal case: theories become clear and 'reasonable' only *after* incoherent parts of them have been used for a long time. Such unreasonable, nonsensical, unmethodical foreplay thus turns out to be an unavoidable precondition of clarity and of empirical success.

Now, when we attempt to describe and to understand developments of this kind in a general way, we are, of course, obliged to appeal to the existing forms of speech which do not take them into account and which must be distorted, misused, beaten into new patterns in order to fit unforeseen situations (without a constant misuse of language there can not be any discovery, any progress). 'Moreover, since the traditional categories are the gospel of everyday thinking (including ordinary scientific thinking) and of everyday practice, [such an attempt at understanding] in effect presents rules and forms of false thinking and action – false, that is, from the standpoint of (scientific) common sense.'[2] This is how *dialectical thinking* arises as a form of thought that 'dissolves into nothing the detailed determinations of the understanding',[3] formal logic included.

(Incidentally, it should be pointed out that my frequent use of such words as 'progress', 'advance', 'improvement', etc., does not mean that I claim to possess special knowledge about what is good and what is bad in the sciences and that I want to impose this knowledge upon my readers. *Everyone can read the terms in his own way* and in accordance with the tradition to which he belongs. Thus for an empiricist, 'progress' will mean transition to a theory that provides direct empirical tests for most of its basic assumptions. Some people believe the quantum theory to be a theory of this kind. For others, 'progress' may mean unification and harmony, perhaps even at the expense of empirical adequacy. This is how Einstein viewed the general theory of relativity. *And my thesis is that anarchism helps to achieve progress in any one of the senses one cares to choose.* Even a law-and-order science will succeed only if anarchistic moves are occasionally allowed to take place.)

It is clear, then, that the idea of a fixed method, or of a fixed theory of rationality, rests on too naive a view of man and his social surroundings. To those who look at the rich material provided by history, and who are not intent on impoverishing it in order to please their lower instincts, their craving for intellectual security in the form of clarity,

2. Herbert Marcuse, *Reason and Revolution*, London, 1941, p. 130.
3. Hegel, *Wissenschaft der Logik*, Vol. 1, Meiner, Hamburg, 1965, p. 6.

precision, 'objectivity', 'truth', it will become clear that there is only *one* principle that can be defended under *all* circumstances and in *all* stages of human development. It is the principle: *anything goes*.

This abstract principle must now be examined and explained in concrete detail.

2

For example, we may use hypotheses that contradict well-confirmed theories and/or well-established experimental results. We may advance science by proceeding counterinductively.

Examining the principle in concrete detail means tracing the consequences of 'counterrules' which oppose some familiar rules of the scientific enterprise. To see how this works, let us consider the rule that it is 'experience', or the 'facts', or 'experimental results' which measure the success of our theories, that agreement between a theory and the 'data' favours the theory (or leaves the situation unchanged) while disagreement endangers it, and perhaps even forces us to eliminate it. This rule is an important part of all theories of confirmation and corroboration. It is the essence of empiricism. The 'counterrule' corresponding to it advises us to introduce and elaborate hypotheses which are inconsistent with well-established theories and/or well-established facts. It advises us to proceed *counterinductively*.

The counterinductive procedure gives rise to the following questions: Is counterinduction more reasonable than induction? Are there circumstances favouring its use? What are the arguments for it? What are the arguments against it? Is perhaps induction always preferable to counterinduction? And so on.

These questions will be answered in two steps. I shall first examine the counterrule that urges us to develop hypotheses inconsistent with accepted and highly confirmed *theories*. Later on I shall examine the counterrule that urges us to develop hypotheses inconsistent with well-established *facts*. The results may be summarized as follows.

In the first case it emerges that the evidence that might refute a theory can often be unearthed only with the help of an incompatible alternative: the advice (which goes back to Newton and which is still very popular today) to use alternatives only when refutations have already discredited the orthodox theory puts the cart before the horse. Also, some of the most important formal properties of a theory are found by contrast, and not by analysis. A scientist who wishes to maximize the empirical content of the views he holds and who wants to understand them as clearly as he possibly can must therefore introduce other views; that is, he must adopt a *pluralistic methodology*. He must compare ideas with other ideas rather than with 'experience' and he must try to improve rather than discard the views that have failed in the competition. Proceeding in this way he will retain the theories of man and cosmos that are found in Genesis, or in the Pimander, he will elaborate them and use them to measure the success of evolution and other 'modern' views.[1] He may then discover that the theory of evolution is not as good as is generally assumed and that it must be supplemented, or entirely replaced, by an improved version of Genesis. Knowledge so conceived is not a series of self-consistent theories that converges towards an ideal view; it is not a gradual approach to the truth. It is rather an ever increasing *ocean of mutually incompatible (and perhaps even incommensurable) alternatives*, each single theory, each fairy tale, each myth that is part of the collection forcing the others into greater articulation and all of them contributing, via this process of competition, to the development of our consciousness. Nothing is ever settled, no view can ever be omitted from a comprehensive account. Plutarch, or Diogenes Laertius and not Dirac, or von Neumann are the models for presenting a knowledge of this kind in which the *history* of a science becomes an inseparable part of the science itself – it is essential for its further *development* as well as for giving *content* to the theories it contains at any particular moment. Experts and laymen, professionals and dilettanti, truth-freaks and liars – they all are invited to participate in the contest and to make their contribution to the enrichment of our culture. The task of the scientist, however, is no longer 'to search for the truth', or 'to praise god', or 'to systematize observations', or 'to improve predictions'. These are but side effects of an activity to which his attention is now mainly directed and which is '*to make the weaker case the stronger*' as the sophists said, *and thereby to sustain the motion of the whole*.

The second 'counterrule' which favours hypotheses inconsistent with

1. For the role of the Pimander in the Copernican Revolution cf. note 12 to Chapter 8.

observations, facts and experimental results, needs no special defence, for there is not a single interesting theory that agrees with all the known facts in its domain. The question is, therefore, not whether counter-inductive theories should be *admitted* into science; the question is, rather, whether the *existing* discrepancies between theory and fact should be increased, or diminished, or what else should be done with them.

To answer this question it suffices to remember that observational reports, experimental results, 'factual' statements, either *contain* theoretical assumptions or *assert* them by the manner in which they are used. (For this point cf. the discussion of natural interpretations in Chapters 6ff.) Thus our habit of saying 'the table is brown' when we view it under normal circumstances, with our senses in good order, but 'the table seems to be brown' when either the lighting conditions are poor or when we feel unsure in our capacity of observation expresses the belief that there are familiar circumstances when our senses are capable of seeing the world 'as it really is' and other, equally familiar circumstances, when they are deceived. It expresses the belief that some of our sensory impressions are veridical while others are not. We also take it for granted that the material medium between the object and us exerts no distorting influence, and that the physical entity that establishes the contact – light – carries a true picture. All these are abstract, and highly doubtful, assumptions which shape our view of the world without being accessible to a direct criticism. Usually, we are not even aware of them and we recognize their effects only when we encounter an entirely different cosmology: prejudices are found by contrast, not by analysis. The material which the *scientist* has at his disposal, his most sublime theories and his most sophisticated techniques included, is structured in exactly the same way. It again contains principles which are not known and which, if known, would be extremely hard to test. (As a result, a theory may clash with the evidence not because it is not correct, but because the evidence is contaminated.)

Now – how can we possibly examine something we are using all the time? How can we analyse the terms in which we habitually express our most simple and straightforward observations, and reveal their presuppositions? How can we discover the kind of world we presuppose when proceeding as we do?

The answer is clear: we cannot discover it from the *inside*. We need an *external* standard of criticism, we need a set of alternative assumptions or, as these assumptions will be quite general, constituting, as it were, an entire alternative world, *we need a dream-world in order to discover the features of the real world we think we inhabit* (and which may actually be just another dream-world). The first step in our criticism of familiar concepts and procedures, the first step in our criticism of 'facts', must therefore be an attempt to break the circle. We must invent a new conceptual system that suspends, or clashes with the most carefully established observational results, confounds the most plausible theoretical principles, and introduces perceptions that cannot form part of the existing perceptual world.[2] This step is again counterinductive. Counterinduction is therefore, always reasonable and it has always a chance of success.

In the following seven chapters, this conclusion will be developed in greater detail and it will be elucidated with the help of historical examples. One might therefore get the impression that I recommend a new methodology which replaces induction by counterinduction and uses a multiplicity of theories, metaphysical views, fairy-tales instead of the customary pair theory/observation.[3] This impression would certainly be mistaken. My intention is not to replace one set of general rules by another such set: my intention is, rather, to convince the reader that *all methodologies, even the most obvious ones, have their limits*. The best way to show this is to demonstrate the limits and even the irrationality of some rules which she, or he, is likely to regard as basic. In the case of induction (including induction by falsification) this means demonstrating how well the counterinductive procedure can be supported by argument. Always remember that the demonstrations and the rhetorics used do not express any 'deep convictions' of mine. They merely show how easy it is to lead people by the nose in a rational way. An anarchist is like

2. 'Clashes' or 'suspends' is meant to be more general than 'contradicts'. I shall say that a set of ideas or actions 'clashes' with a conceptual system if it is either inconsistent with it, or makes the system appear absurd. For details cf. Chapter 17 below.

3. This is how Professor Ernan McMullin interpreted some earlier papers of mine. See 'A Taxonomy of the Relations between History and Philosophy of Science', *Minnesota Studies* 5, Minneapolis, 1971.

an undercover agent who plays the game of Reason in order to undercut the authority of Reason (Truth, Honesty, Justice, and so on).[4]

4. 'Dada', says Hans Richter in *Dada: Art and Anti-Art*, 'not only had no programme, it was against all programmes.' This does not exclude the skilful defence of programmes to show the chimerical character of any defence, however 'rational'. Cf. also Chapter 16, text to footnotes, 21, 22, 23. (In the same way an actor or a playwright could produce all the outer manifestations of 'deep love' in order to debunk the idea of 'deep love' itself. Example: Pirandello.) These remarks, I hope, will alleviate Miss Koertge's fear that I intend to start just another movement, the slogans 'proliferate' or 'anything goes' replacing the slogans of falsificationism or inductivism or research-programmism.

3

The consistency condition which demands that new hypotheses agree with accepted theories *is unreasonable because it preserves the older theory, and not the better theory. Hypotheses contradicting well-confirmed theories give us evidence that cannot be obtained in any other way. Proliferation of theories is beneficial for science, while uniformity impairs its critical power. Uniformity also endangers the free development of the individual.*

In this chapter I shall present more detailed arguments for the 'counter-rule' that urges us to introduce hypotheses which are *inconsistent* with well-established *theories.* The arguments will be indirect. They will start with a criticism of the demand that new hypotheses must be *consistent* with such theories. This demand will be called the *consistency condition.*[1]

Prima facie, the case of consistency condition can be dealt with in a few words. It is well known (and has also been shown in detail by Duhem) that Newton's theory is inconsistent with Galileo's law of free fall and with Kepler's laws; that statistical thermodynamics is inconsistent with the second law of the phenomenological theory; that wave optics is inconsistent with geometrical optics; and so on.[2] Note that

1. The consistency condition goes back to Aristotle at least. It plays an important part in Newton's philosophy (though Newton himself constantly violated it). It is taken for granted by the majority of 20th-century philosophers of science.

2. Pierre Duhem, *La Théorie Physique : Son Objet, Sa Structure*, Paris, 1914, Chapters IX and X. In his *Objective Knowledge*, Oxford, 1972, pp. 204f, Karl Popper quotes me in support of his contention that he originated the idea 'that theories may *correct* an "observational" or "phenomenal" law which they are supposed to explain'. He makes two mistakes. The first mistake is that he takes my references to him as historical evidence for his priority while they are just friendly gestures. The second mistake is that the idea quoted occurs in Duhem, Einstein and especially in Boltzmann who anticipated every philosophical observation of 'The Aim of Science', *Ratio* i, pp. 24ff,

what is being asserted here is *logical* inconsistency; it may well be that the differences of prediction are too small to be detected by experiment. Note also that what is being asserted is not the *inconsistency* of, say, Newton's *theory* and Galileo's law, but rather the inconsistency of *some consequences* of Newton's theory in the domain of validity of Galileo's law, and Galileo's law. In the last case, the situation is especially clear. Galileo's law asserts that the acceleration of free fall is a constant, whereas application of Newton's theory to the surface of the earth gives an acceleration that is not constant but *decreases* (although imperceptibly) with the distance from the centre of the earth.

To speak more abstractly: consider a theory T′ that successfully describes the situation inside domain D′. T′ agrees with a *finite* number of observations (let their class be F) and it agrees with these observations inside a margin M of error. Any alternative that contradicts T′ outside F and inside M is supported by exactly the same observations and is therefore acceptable if T′ was acceptable (I shall assume that F are the only observations made). The consistency condition is much less tolerant. It eliminates a theory or a hypothesis not because it disagrees with the facts; it eliminates it because it disagrees with another theory, with a theory, moreover, whose confirming instances it shares. It thereby makes the as yet untested part of that theory a measure of validity. The only difference between such a measure and a more recent theory is age and familiarity. Had the younger theory been there first, then the consistency condition would have worked in its favour. 'The *first* adequate theory has the right of priority over equally adequate aftercomers.'[3] In this respect the effect of the consistency condition is rather similar to the effect of the more traditional methods of transcendental deduction, analysis of essences, phenomenological analysis, linguistic analysis. It contributes to the preservation of the old and familiar not because of any inherent advantage in it – for example, not because it has a better foundation in observation than has the newly suggested alternative, or because it is more elegant – but because it is old and familiar. This is not the only

---

and of its predecessors. For Boltzmann cf. my article in the *Encyclopaedia of Philosophy*, ed. Paul Edwards. For Duhem cf. *Objective Knowledge*, p. 200.

3. C. Truesdell, 'A Program Toward Rediscovering the Rational Mechanics of the Age of Reason', *Archives for the History of Exact Sciences*, Vol. 1, p. 14.

instance where on closer inspection a rather surprising similarity emerges between modern empiricism and some of the school philosophies it attacks.

Now it seems to me that these brief considerations, although leading to an interesting *tactical* criticism of the consistency condition, and to some first shreds of support for counterinduction, do not yet go to the heart of the matter. They show that an alternative to the accepted point of view which shares its confirming instances cannot be *eliminated* by factual reasoning. They do not show that such an alternative is *acceptable*; and even less do they show that it *should be used*. It is bad enough, a defender of the consistency condition might point out, that the accepted view does not possess full empirical support. Adding new theories *of an equally unsatisfactory character* will not improve the situation; nor is there much sense in trying to *replace* the accepted theories by some of their possible alternatives. Such replacement will be no easy matter. A new formalism may have to be learned and familiar problems may have to be calculated in a new way. Textbooks must be rewritten, university curricula readjusted, experimental results reinterpreted. And what will be the result of all the effort? Another theory which, from an empirical standpoint has no advantage whatsoever over and above the theory it replaces. The only real improvement, so the defender of the consistency condition will continue, derives from the *addition of new facts*. Such new facts will either support the current theories, or they will force us to modify them by indicating precisely where they go wrong. In both cases they will precipitate real progress and not merely arbitrary change. The proper procedure must therefore consist in the confrontation of the accepted point of view with as many relevant facts as possible. The exclusion of alternatives is then simply a measure of expediency: their invention not only does not help, it even hinders progress by absorbing time and manpower that could be devoted to better things. The consistency condition eliminates such fruitless discussion and it forces the scientist to concentrate on the facts which, after all, are the only acceptable judges of a theory. This is how the practising scientist will defend his concentration on a single theory to the exclusion of empirically possible alternatives.[4]

4. More detailed evidence for the existence of this attitude and its influence on the development of the sciences may be found in Thomas Kuhn, *The Structure of Scientific*

It is worthwhile repeating the reasonable core of this argument. Theories should not be changed unless there are pressing reasons for doing so. The only pressing reason for changing a theory is disagreement with facts. Discussion of incompatible facts will therefore lead to progress. Discussion of incompatible hypotheses will not. Hence, it is sound procedure to increase the number of relevant facts. It is not sound procedure to increase the number of factually adequate, but incompatible, alternatives. One might wish to add that formal improvements such as increased elegance, simplicity, generality, and coherence should not be excluded. But once these improvements have been carried out, the collection of facts for the purpose of tests seems indeed to be the only thing left to the scientist.

And so it is – provided facts *exist, and are available independently of whether or not one considers alternatives to the theory to be tested.* This assumption, on which the validity of the foregoing argument depends in a most decisive manner, I shall call the assumption of the relative autonomy of facts, or the *autonomy principle.* It is not asserted by this principle that the discovery and description of facts is independent of *all* theorizing. But it *is* asserted that the facts which belong to the empirical content of some theory are available whether or not one considers alternatives to *this* theory. I am not aware that this very important assumption has ever been explicitly formulated as a separate postulate of the empirical method. However, it is clearly implied in almost all investigations which deal with questions of confirmation and test. All these investigations use a model in which a *single* theory is compared with a class of facts (or observation statements) which are assumed to be 'given' somehow. I submit that this is much too simple a picture of the actual

---

Revolutions, Chicago, 1962. The attitude is extremely common in the quantum theory. 'Let us enjoy the successful theories we possess and let us not waste our time with contemplating what *would* happen if *other* theories were used' seems to be the guiding philosophy of almost all contemporary physicists (cf. for example, W. Heisenberg, *Physics and Philosophy*, New York, 1958, pp. 56 and 144) and 'scientific' philosophers (e.g. N. R. Hanson, 'Five Cautions for the Copenhagen Critics', *Philosophy of Science*, No. 26, 1959, pp. 325ff). It may be traced back to Newton's papers and letters (to Hooke, Pardies, and others) on the theory of colours and to his general methodology (cf. my account in 'Classical Empiricism', *The Methodological Heritage of Newton*, ed. Butts, Oxford, 1970).

situation. Facts and theories are much more intimately connected than is admitted by the autonomy principle. Not only is the description of every single fact dependent on *some* theory (which may, of course, be very different from the theory to be tested), but there also exist facts which cannot be unearthed except with the help of alternatives to the theory to be tested, and which become unavailable as soon as such alternatives are excluded. This suggests that the methodological unit to which we must refer when discussing questions of test and empirical content is constituted by a *whole set of partly overlapping, factually adequate, but mutually inconsistent theories*. In the present chapter only the barest outlines will be given of such a test model. However, before doing this, I want to discuss an example which shows very clearly the function of alternatives in the discovery of critical facts.

It is now known that the Brownian particle is a perpetual motion machine of the second kind and that its existence refutes the phenomenological second law. Brownian motion therefore belongs to the domain of relevant facts for the law. Now could this relation between Brownian motion and the law have been discovered in a *direct* manner i.e. could it have been discovered by an examination of the observational consequences of the phenomenological theory that did not make use of an alternative theory of heat? This question is readily divided into two: (1) Could the *relevance* of the Brownian particle have been discovered in this manner? (2) Could it have been demonstrated that it actually *refutes* the second law?

The answer to the first question is that we do not know. It is impossible to say what would have happened if the kinetic theory had not been introduced into the debate. It is my guess, however, that in that case the Brownian particle would have been regarded as an oddity – in much the same way as some of the late Professor Ehrenhaft's astounding effects were regarded as an oddity,[5] and that it would not have been given the

5. Having witnessed these phenomena under a great variety of conditions I am far more reluctant to dismiss them as a mere *Dreckeffekt* than the scientific community of today. Cf. my translation of Ehrenhaft's Vienna lectures of 1947 which can be obtained from me at the drop of a postcard. Ehrenhaft was regarded by many of his colleagues as a charlatan. If he was, he was a far better teacher than most of them and he gave his students a much better idea of the precarious character of physical knowledge. I still remember how eagerly we studied Maxwell's theory (from the Abraham–Becker text-

decisive position it assumes in contemporary theory. The answer to the second question is simply – No. Consider what the discovery of an inconsistency between the phenomenon of Brownian motion and the second law would have required. It would have required: (a) measurement of the exact *motion* of the particle in order to ascertain the change in its kinetic energy plus the energy spent on overcoming the resistance of the fluid; and (b) it would have required precise measurements of temperature and heat transfer in the surrounding medium in order to establish that any loss occurring there was indeed compensated by the increase in the energy of the moving particle and the work done against the fluid. Such measurements are beyond experimental possibilities:[6] neither the heat transfer nor the path of the particle can be measured with the desired precision. Hence a 'direct' refutation of the second law that would consider only the phenomenological theory and the 'facts' of the Brownian motion, is impossible. It is impossible because of the structure of the world in which we live and because of the laws that are valid in this world. And as is well known, the actual refutation was brought about in a very different manner. It was brought about via the kinetic theory and Einstein's utilization of it in his calculation of the statistical properties of Brownian motion. In the course of this procedure, the phenomenological theory (T′) was incorporated into the wider context of statistical physics (T) in such a manner that *the consistency condition was violated*, and it was only *then* that a crucial experiment was staged (investigations of Svedberg and Perrin).[7]

---

book, from Heaviside whom Ehrenhaft frequently mentioned in his lectures, and from Maxwell's original papers) and the relativity theory, in order to refute his claim that theoretical physics was just nonsense; and how astonished and disappointed we were to discover that there was no straightforward deductive chain from theory to experiment and that many published derivations were quite arbitrary. We also realized that almost all theories derive their strength from a few paradigmatic cases and that they have to be distorted in order to cope with the rest. It is a pity that philosophers of science only rarely take up boundary cases like Ehrenhaft or Velikovsky and that they prefer recognition by the top dogs in science (and in their own disaster area) to increased insight into the scientific enterprise.

6. For details cf. R. Furth, *Zs. Physik*, Vol. 81 (1933), pp. 143ff.

7. For these investigations (whose philosophical background derives from Boltzmann) cf. A. Einstein, *Investigations on the Theory of the Brownian Motion*, ed. R. Fürth, New York, 1956, which contains all the relevant papers by Einstein and an exhaustive bib-

It seems to me that this example is typical of the relation between fairly general theories, or points of view, and the 'facts'. Both the relevance and the refuting character of decisive facts can be established only with the help of other theories which, though factually adequate,[8] are not in agreement with the view to be tested. This being the case, the invention and articulation of alternatives may have to precede the production of refuting facts. Empiricism, at least in some of its more sophisticated versions, demands that the empirical content of whatever knowledge we possess be increased as much as possible. *Hence the invention of alternatives to the view at the centre of discussion constitutes an essential part of the empirical method.* Conversely the fact that the consistency condition eliminates alternatives now shows it to be in disagreement not only with scientific practice but with empiricism as well. By excluding valuable tests it decreases the empirical content of the theories that are permitted to remain (and these, as I have indicated above, will usually be the theories which were there first); and it especially decreases the number of those facts that could show their limitations. This last result of a determined application of the consistency condition

---

liography by R. Fürth. For the experimental work of J. Perrin, see *Die Atome*, Leipzig, 1920. For the relation between the phenomenological theory and the kinetic theory of von Smoluchowski, see 'Experimentell nachweisbare, der üblichen Thermodynamik widersprechende Molekularphänomene', *Physikalische Zs.*, xiii, 1912, p. 1069, as well as the brief note by K. R. Popper, 'Irreversibility, or, Entropy since 1905', *British Journal for the Philosophy of Science*, viii, 1957, p. 151, which summarizes the essential arguments. Despite Einstein's epoch-making discoveries and von Smoluchowski's splendid presentation of their consequences (*Oeuvres de Marie Smoluchowski*, Cracovie, 1927, Vol. ii, pp. 226ff, 316ff, 462ff and 530ff), the present situation in thermodynamics is extremely unclear, especially in view of the continued presence of some very doubtful ideas of reduction. To be more specific, the attempt is frequently made to determine the entropy balance of a complex *statistical* process by reference to the (refuted) *phenomenological* law after which fluctuations are inserted in an *ad hoc* fashion. For this cf. my note 'On the Possibility of a Perpetuum Mobile of the Second Kind', *Mind, Matter and Method*, Minneapolis, 1966, p. 409, and my paper 'In Defence of Classical Physics', *Studies in the History and Philosophy of Science*, 1, No. 2, 1970.

It ought to be mentioned, incidentally, that in 1903, when Einstein started his work in thermodynamics, there existed empirical evidence suggesting that Brownian motion could not be a molecular phenomenon. See F. M. Exner, 'Notiz zu Browns Molekularbewegung', *Ann. Phys.*, No. 2, 1900, p. 843. Exner claimed that the motion was of orders of magnitude beneath the value to be expected on the equipartition principle.

8. The condition of factual adequacy will be removed in Chapter 5.

is of very topical interest. It may well be that the refutation of the quantum-mechanical uncertainties presupposes just such an incorporation of the present theory into a wider context which no longer agrees with the idea of complementarity and which therefore suggests new and decisive experiments. And it may also be that the insistence, on the part of the majority of contemporary physicists, on the consistency condition will, if successful, forever protect the uncertainties from refutation. This is how the condition may finally create a situation where a certain point of view petrifies into dogma by being, in the name of experience, completely removed from any conceivable criticism.

It is worthwhile examining this apparently 'empirical' defence of a dogmatic point of view in somewhat greater detail. Assume that physicists have adopted, either consciously or unconsciously, the idea of the uniqueness of complementarity and that they elaborate the orthodox point of view and refuse to consider alternatives. In the beginning such a procedure may be quite harmless. After all, a man and even an influential school can do only so many things at a time and it is better when they pursue a theory in which they are interested rather than a theory they find boring. Now assume that the pursuit of the chosen theory has led to successes, and that the theory has explained, in a satisfactory manner, circumstances that had been unintelligible for quite some time. This gives empirical support to an idea which to start with seemed to possess only this advantage: it was interesting and intriguing. The commitment to the theory will now be reinforced, and the attitude towards alternatives will become less tolerant. Now if it is true, as has been argued in the last section, that many facts become available only with the help of alternatives, then the refusal to consider them *will result in the elimination of potentially refuting facts as well.* More especially, it will eliminate facts whose discovery would show the complete and irreparable inadequacy of the theory.[9] Such facts having been made inaccessible, the

9. The quantum theory can be adapted to a great many difficulties. It is an open theory, in the sense that apparent inadequacies can be accounted for in an *ad hoc* manner, by adding suitable operators, or elements in the Hamiltonian, rather than by recasting the whole structure. A refutation of the basic formalism would, therefore, have to prove *that there is no conceivable adjustment of the Hamiltonian, or of the operators used*, that could make the theory conform to a given fact. Clearly such a general statement can only be provided by an *alternative* theory which must be detailed enough to

theory will appear to be free from blemish and it will seem that 'all evidence points with merciless definiteness in the . . . direction . . . that all the processes involving . . . unknown interactions conform to the fundamental quantum law'.[10] This will further reinforce the belief in the uniqueness of the accepted theory and in the futility of any account that proceeds in a different manner. Being now very firmly convinced that there is only one good microphysics, the physicists will try to explain adverse facts in its terms, and they will not mind when such explanations occasionally turn out to be a little clumsy. Next, the development becomes known to the public. Popular science books (and this includes many books on the philosophy of science) spread the basic postulates of the theory; applications are made in distant fields, money is given to the orthodox, and is withheld from the rebels. More than ever the theory seems to possess tremendous empirical support. The chances for the consideration of alternatives are now very slight indeed. The final success of the fundamental assumptions of the quantum theory, and of the idea of complementarity, seems to be assured.

At the same time it is evident, on the basis of our considerations, that this appearance of success *cannot in the least be regarded as a sign of truth and correspondence with nature.* Quite the contrary, the suspicion arises that the absence of major difficulties is a result of the decrease of empirical content brought about by the elimination of alternatives, and of facts that can be discovered with their help. In other words, *the suspicion arises that this alleged success is due to the fact that the theory, when extended*

---

allow for crucial tests. This has been explained by D. Bohm and J. Bub, *Reviews of Modern Physics*, No. 38, 1966, pp. 456ff. The observations which refute a theory are not always *discovered* with the help of an alternative, they are often known in advance. Thus the anomaly of the perihelion of Mercury was known long before the invention of the general theory of relativity (which in turn was not invented to solve this problem). The Brownian particle was known long before the more detailed versions of the kinetic theory were available. But their explanation with the help of an alternative certainly makes us see them in a new light: we now find out that they conflict with a generally accepted view. I have the suspicion that all 'falsifications', including even the trite Case of the White Raven (or the Black Swan) are based on discoveries of the latter kind. For a most interesting discussion of the notion of 'novelty' that arises in this connection see section 1.1 of Elie Zahar's 'Why Did Einstein's Programme supersede Lorentz's?', *British Journal for the Philosophy of Science*, June 1973.

10. L. Rosenfeld, 'Misunderstandings about the Foundations of the Quantum Theory', *Observation and Interpretation*, ed. Körner, London, 1957, p. 44.

beyond its starting point, was turned into rigid ideology. Such ideology is 'successful' not because it agrees so well with the facts; it is successful because no facts have been specified that could constitute a test, and because some such facts have even been removed. Its 'success' *is entirely man-made*. It was decided to stick to some ideas, come what may, and the result was, quite naturally, the survival of these ideas. If now the initial decision is forgotten, or made only implicitly, for example, if it becomes common law in physics, then the survival itself will seem to constitute independent support, it will reinforce the decision, or turn it into an explicit one, and in this way close the circle. This is how empirical 'evidence' may be *created* by a procedure which quotes as its justification the very same evidence it has produced.

At this point an 'empirical' theory of the kind described (and let us always remember that the basic principles of the present quantum theory, and especially the idea of complementarity, are uncomfortably close to forming such a theory) becomes almost indistinguishable from a second-rate myth. In order to realize this, we need only consider a myth such as the myth of witchcraft and of demonic possession that was developed by Roman Catholic theologians and that dominated 15th-, 16th- and 17th-century thought on the European continent. This myth is a complex explanatory system that contains numerous auxiliary hypotheses designed to cover special cases, so it easily achieves a high degree of confirmation on the basis of observation. It has been taught for a long time; its content is enforced by fear, prejudice, and ignorance, as well as by a jealous and cruel priesthood. Its ideas penetrate the most common idiom, infect all modes of thinking and many decisions which mean a great deal in human life. It provides models for the explanation of any conceivable event – conceivable, that is, for those who have accepted it.[11] This being the case, its key terms will be fixed in an unambiguous manner and the idea (which may have led to such a procedure in the first place) that they are copies of unchanging entities and that change of meaning, if it should happen, is due to human mistake – this idea will now be very plausible. Such plausibility reinforces all the manoeuvres which are used for the

11. For detailed descriptions cf. Ch. H. Lea, *Materials for a History of Witchcraft*, New York, 1957, as well as H. Trevor-Roper, *The European Witch Craze*, New York, 1969, which contains plentiful literature, ancient as well as modern.

preservation of the myth (elimination of opponents included). The conceptual apparatus of the theory and the emotions connected with its application, having penetrated all means of communication, all actions, and indeed the whole life of the community, now guarantees the success of methods such as transcendental deduction, analysis of usage, phenomenological analysis – which are means for further solidifying the myth (which shows, by the way, that all these methods, which have been the trade mark of various philosophical schools old and new, have one thing in common: they tend to *preserve the status quo* of intellectual life). Observational results, too, will speak in favour of the theory, as they are formulated in its terms. It will seem that the truth has at last been arrived at. At the same time, it is evident that all contact with the world has been lost and that the stability achieved, the semblance of absolute truth, *is nothing but the result of an absolute conformism.*[12] For how can we possibly test, or improve upon, the truth of a theory if it is built in such a manner that any conceivable event can be described, and explained, in terms of its principles? The *only* way of investigating such all-embracing principles would be to compare them with a different set of *equally all-embracing principles* – but this procedure has been excluded from the very beginning. The myth is, therefore, of no objective relevance; it continues to exist solely as the result of the effort of the community of believers and of their leaders, be these now priests or Nobel prize winners. This, I think, is the most decisive argument against any method that encourages uniformity, be it empirical or not. Any such method is, in the last resort, a method of deception. It enforces an unenlightened conformism, and speaks of truth; it leads to a deterioration of intellectual capabilities, of the power of imagination, and speaks of deep insight; it destroys the most precious gift of the young – their tremendous power of imagination, and speaks of education.

12. Analysis of usage, to take only one example, presupposes certain regularities concerning this usage. The more people differ in their fundamental ideas, the more difficult it will be to uncover such regularities. Hence, analysis of usage will work best in a closed society that is firmly held together by a powerful myth such as was the society of Oxford philosophers of about 20 years ago. – Schizophrenics very often hold beliefs which are as rigid, all-pervasive, and unconnected with reality, as are the best dogmatic philosophies. However, such beliefs come to them naturally whereas a 'critical' philosopher may sometimes spend his whole life in attempting to find arguments which create a similar state of mind.

To sum up: *Unanimity of opinion may be fitting for a church, for the frightened or greedy victims of some (ancient, or modern) myth, or for the weak and willing followers of some tyrant. Variety of opinion is necessary for objective knowledge. And a method that encourages variety is also the only method that is compatible with a humanitarian outlook.* (To the extent to which the consistency condition delimits variety, it contains a theological element which lies, of course, in the worship of 'facts' so characteristic of nearly all empiricism.[13])

13. It is interesting to see that the platitudes that directed the Protestants to the Bible are often almost identical with the platitudes which direct empiricists and other fundamentalists to *their* foundation, viz. experience. Thus in his *Novum Organum* Bacon demands that all preconceived notions (aphorism 36), opinions (aphorisms 42ff), even *words* (aphorisms 59, 121), 'be adjured and renounced with firm and solemn resolution, and the understanding must be completely freed and cleared of them, so that the access to the kingdom of man, which is founded on the sciences, may resemble that to the kingdom of heaven, where no admission is conceded, except to children' (aphorism 68). In both cases 'disputation' (which is the consideration of alternatives) is criticized, in both cases we are invited to dispense with it, and in both cases we are promised an 'immediate perception', here, of God, and there of Nature. For the theoretical background of this similarity cf. my essay 'Classical Empiricism', in *The Methodological Heritage of Newton*, ed. R. E. Butts, Oxford and Toronto, 1970. For the strong connections between Puritanism and modern science see R. T. Jones, *Ancients and Moderns*, California, 1965, Chapters 5–7. A thorough examination of the numerous factors that influenced the rise of modern empiricism in England is found in R. K. Merton, *Science, Technology and Society in Seventeenth Century England*, New York, Howard Fertig, 1970 (book version of the 1938 article).

4

There is no idea, however ancient and absurd that is not capable of improving our knowledge. The whole history of thought is absorbed into science and is used for improving every single theory. Nor is political interference rejected. It may be needed to overcome the chauvinism of science that resists alternatives to the status quo.

This finishes the discussion of part one of counterinduction dealing with the invention and elaboration of hypotheses inconsistent with a point of view that is highly confirmed and generally accepted. It was pointed out that the examination of such a point of view often needs an incompatible alternative theory so that the (Newtonian) advice to postpone alternatives until after the first difficulty has arisen means putting the cart before the horse. A scientist who is interested in maximal empirical content, and who wants to understand as many aspects of his theory as possible, will accordingly adopt a pluralistic methodology, he will compare theories with other theories rather than with 'experience', 'data', or 'facts', and he will try to improve rather than discard the views that appear to lose in the competition.[1] For the alternatives, which he needs to keep the contest going, may be taken from the past as well. As a matter of fact, they may be taken from wherever one is able to find them – from ancient myths and modern prejudices; from the lucubrations of experts and from the fantasies of cranks. The whole history of a subject is utilized in the attempt to improve its most recent and most 'advanced' stage. The separation between the history of a science, its philosophy and the

1. It is, therefore, important that the alternatives be set against each other and not be isolated or emasculated by some form of 'demythologisation'. Unlike Tillich, Bultmann and their followers, we should regard the world-views of the Bible, the Gilgamesh epic, the Iliad, the Edda, as fully fledged *alternative cosmologies* which can be used to modify, and even to replace, the 'scientific' cosmologies of a given period.

science itself dissolves into thin air and so does the separation between science and non-science.[2]

This position, which is a natural consequence of the arguments presented above, is frequently attacked – not by counter-arguments, which would be easy to answer, but by rhetorical questions. 'If any metaphysics goes,' writes Dr Hesse in her review of an earlier essay of mine,[3] 'then the question arises why we do not *go back* and exploit the objective criticism of modern science available in Aristotelianism, or indeed in Voodoo?' – and she insinuates that a criticism of this kind would be altogether laughable. Her insinuation, unfortunately, assumes a great deal of ignorance in her readers. Progress was often achieved by

2. An account and a truly humanitarian defence of this position can be found in J. S. Mill's *On Liberty*. Popper's philosophy, which some people would like to lay on us as the one and only humanitarian rationalism in existence today, is but a pale reflection of Mill. It is much more specialized, much more formalistic and elitist, and quite devoid of the concern for individual happiness that is such a characteristic feature of Mill. We can understand its peculiarities when we consider (a) the background of logical positivism, which plays an important role in the *Logic of Scientific Discovery*, (b) the unrelenting puritanism of its author (and of most of his followers), and when we remember the influence of Harriet Taylor on Mill's life and on his philosophy. There is no Harriet Taylor in Popper's life. The foregoing arguments should also have made it clear that I regard proliferation not just as an 'external catalyst' of progress, as Lakatos suggests in his essays ('History of Science and its Rational Reconstructions', *Boston Studies*, Vol. VIII, p. 98; 'Popper on Demarcation and Induction', MS, 1970, p. 21), but as an essential part of it. Ever since 'Explanation, Reduction, and Empiricism' (*Minnesota Studies*, Vol. III, Minneapolis, 1962), and especially in 'How to be a Good Empiricist' (*Delaware Studies*, Vol. II, 1963), I have argued that alternatives increase the empirical content of the views that happen to stand in the centre of attention and are, therefore, '*necessary* parts' of the falsifying process (Lakatos, *History*, fn. 27, describing his own position). In 'Reply to Criticism' (*Boston Studies*, Vol. II, 1965) I pointed out that 'the principle of proliferation not only recommends invention of *new* alternatives, it also prevents the elimination of *older* theories which have been refuted. The reason is that such theories contribute to the content of their victorious rivals' (p. 224). This agrees with Lakatos' observation of 1971 that 'alternatives are not merely catalysts, which can later be removed in the rational reconstruction' (*History*, fn. 27), *except* that Lakatos attributes the psychologistic view to me and my *actual* views to himself. Considering the argument in the text, it is clear that the increasing separation of the history, the philosophy of science and of science itself is a disadvantage and should be terminated in the interest of all these three disciplines. Otherwise we shall get tons of minute, precise, but utterly barren results.

3. Mary Hesse, *Ratio*, No. 9, 1967, p. 93; cf. B. F. Skinner, *Beyond Freedom and Dignity*, New York, 1971, p. 5: 'No modern physicist would turn to Aristotle for help.' This may be true, but is hardly an advantage.

a 'criticism from the past', of precisely the kind that is now dismissed by her. After Aristotle and Ptolemy, the idea that the earth moves – that strange, ancient, and 'entirely ridiculous',[4] Pythagorean view – was thrown on the rubbish heap of history, only to be revived by Copernicus and to be forged by him into a weapon for the defeat of its defeaters. The Hermetic writings played an important part in this revival, which is still not sufficiently understood,[5] and they were studied with care by the great Newton himself.[6] Such developments are not surprising. No idea is ever examined in all its ramifications and no view is ever given all the chances it deserves. Theories are abandoned and superseded by more fashionable accounts long before they have had an opportunity to show their virtues. Besides, ancient doctrines and 'primitive' myths appear strange and nonsensical only because their scientific content is either not known, or is distorted by philologists or anthropologists unfamiliar with the simplest physical, medical or astronomical knowledge.[7] Voodoo,

4. Ptolemy, *Syntaxis*, quoted after the translation of Manitius *Des Claudius Ptolemaeus Handbuch der Astronomie*, Vol. I, Leipzig, 1963, p. 18.

5. For a positive evaluation of the role of the Hermetic writings in the Renaissance cf. F. Yates, *Giordano Bruno and the Hermetic Tradition*, London, 1963, and the literature given there. For a criticism of her position cf. the articles by Mary Hesse and Edward Rosen in Vol. V of the *Minnesota Studies for the Philosophy of Science*, ed. Roger Stuewer, Minnesota, 1970; cf. also note 12 to Chapter 8.

6. cf. J. M. Keynes, 'Newton the Man', in *Essays and Sketches in Biography*, New York, 1956, and, in much greater detail, McGuire & Rattansi, 'Newton and the "Pipes of Pan"', *Notes and Records of the Royal Society*, Vol. 21, No. 2, 1966, pp. 108ff.

7. For the scientific content of some myths cf. C. de Santillana, *The Origin of Scientific Thought*, New York, 1961, especially the Prologue. 'We can see then', writes de Santillana, 'how so many myths, fantastic and arbitrary in semblance, of which the Greek tale of the Argonauts is a late offspring, may provide a terminology of image motifs, a kind of code which is beginning to be broken. It was meant to allow those who knew (a) to determine unequivocally the position of given planets in respect to the earth, to the firmament, and to one another; (b) to present what knowledge there was of the fabric of the world in the form of tales about 'how the world began'; there are two reasons why this code was not discovered earlier. The one is the firm conviction of historians of science that science did not start before Greece and that scientific results can only be obtained with the scientific method as it is practised today (and as it was foreshadowed by Greek scientists). The other reason is the astronomical, geological, etc., ignorance of most Assyriologists, Aegyptologists, Old Testament scholars, and so on: the apparent primitivism of many myths is just the reflection of the primitive astronomical, biological, etc., etc., knowledge of their collectors and translators. Since the discoveries of Hawkins, Marshack and others we have to admit the existence of an international palaeolithic astronomy that gave rise to schools, observatories, scientific

Dr Hesse's *pièce de resistance*, is a case in point. Nobody knows it, everybody uses it as a paradigm of backwardness and confusion. And yet Voodoo has a firm though still not sufficiently understood material basis, and a study of its manifestations can be used to enrich, and perhaps even to revise, our knowledge of physiology.[8]

An even more interesting example is the revival of traditional medicine in Communist China. We start with a familiar development:[9] a great country with great traditions is subjected to Western domination and is exploited in the customary way. A new generation recognizes or thinks it recognizes the material and intellectual superiority of the West and traces it back to science. Science is imported, taught, and pushes aside all traditional elements. Scientific chauvinism triumphs: 'What is compatible with science should live, what is not compatible with science, should die'.[10] 'Science' in this context means not just a specific method, but all the results the method has so far produced. Things incompatible with the results must be eliminated. Old style doctors, for example, must either be removed from medical practice, or they must be re-educated. Herbal medicine, acupuncture, moxibustion and the underlying philo-

---

traditions and most interesting theories. These theories which were expressed in sociological, not in mathematical terms, have left their traces in sagas, myths, legends and may be reconstructed in a twofold way, by going *forward* into the present from the material remnants of Stone Age astronomy such as marked stones, stone observatories, etc., and by going *back* into the past from the literary remnants which we find in sagas. An example of the first method is A. Marshack, *The Roots of Civilization*, New York, 1972. An example of the second is de Santillana–von Dechend, *Hamlet's Mill*, Boston, 1969. For a survey and interpretation cf. my *Einführung in die Naturphilosophie*, Braunschweig, 1974.

8. Cf. Chapter 9 of Lévi-Strauss, *Structural Anthropology*, New York, 1967. For the physiological basis of Voodoo cf. C. R. Richter, 'The Phenomenon of Unexplained Sudden Death' in *The Physiological Basis of Psychiatry*, ed. Gantt as well as W. H. Cannon, *Bodily Changes in Pain, Hunger, Fear and Rage*, New York, 1915, and ' "Voodoo" Death', in *American Anthropologist*, n.s., xliv, 1942. The detailed biological and meteorological observations made by so-called 'primitives' are reported in Lévi-Strauss, *The Savage Mind*, London, 1966.

9. R. C. Croizier, *Traditional Medicine in Modern China*, Harvard University Press, 1968. The author gives a very interesting and fair account of developments with numerous quotations from newspapers, books, pamphlets, but is often inhibited by his respect for 20th-century science.

10. Chou Shao, 1933, as quoted in Croizier, op. cit., p. 109. Cf. also D. W. Y. Kwok, *Scientism in Chinese Thought*, New Haven, 1965.

sophy are a thing of the past, no longer to be taken seriously. This was the attitude up to about 1954 when the condemnation of bourgeois elements in the Ministry of Health started a campaign for the revival of traditional medicine. No doubt the campaign was politically inspired. It contained at least two elements, viz. (1) the identification of Western science with bourgeois science and (2) the refusal of the party to except science from political supervision[11] and to grant experts special privileges. But it provided the counterforce that was needed to overcome the scientific chauvinism of the time and to make a plurality (actually a duality) of views possible. (This is an important point. It often happens that parts of science become hardened and intolerant so that proliferation must be enforced from the outside, and by political means. Of course, success cannot be guaranteed – see the Lysenko affair. But this does not remove the need for non-scientific controls on science.)

Now this politically enforced dualism has led to most interesting and puzzling discoveries both in China and in the West and to the realization that there are effects and means of diagnosis which modern medicine cannot repeat and for which it has no explanation.[12] It revealed sizeable lacunae in Western medicine. Nor can one expect that the customary scientific approach will eventually find an answer. In the case of herbal medicine the approach consists of two steps.[13] First, the herbal concoction is analysed into its chemical constituents. Then the *specific* effects of each constituent are determined and the total effect on a particular organ explained on their basis. This neglects the possibility that the herb, taken in its entirety, changes the state of the *whole* organism and that it is

11. For the rationality of this refusal cf. my article 'Experts in a Free Society', *The Critic*, November/December 1970, and Chapter 18 of the present essay. For the tensions between 'red' and 'expert' cf. F. Schurmann, *Ideology and Organization in Communist China*, University of California Press, 1966.

12. For earlier results cf. T. Nakayama, *Acupuncture et Médicine Chinoise Verifiées au Japon*, Paris, 1934, and F. Mann, *Acupuncture*, New York, 1962; revised edition, New York, 1973. In traditional medicine pulse-taking is the chief method of diagnosis, involving 12 different pulses. E. H. Hume, *Doctors East and West*, Baltimore, 1940, pp. 190–2, gives interesting examples where pulse diagnosis and modern scientific diagnosis lead to the same result. Cf. also E. H. Hume, *The Chinese Way of Medicine*, Baltimore, 1940. For the historical background and further material cf. the introduction to *The Yellow Emperor's Classic of Internal Medicine*, transl. Ilza Veith, Berkeley and Los Angeles, 1966.

13. Cf. M. B. Krieg, *Green Medicine*, New York, 1964.

this new state of the whole organism rather than a specific part of the herbal concoction that cures the diseased organ. Here as elsewhere knowledge is obtained from a proliferation of views rather than from the determined application of a preferred ideology. And we realize that proliferation may have to be enforced by non-scientific agencies whose power is sufficient to overcome the most powerful scientific institutions. Examples are the Church, the State, a political party, public discontent, or money: the best single entity to get a modern scientist away from what his 'scientific conscience' tells him to pursue is still the *Dollar* (or, more recently, the German *Mark*).

The examples of Copernicus, the atomic theory, Voodoo, Chinese medicine show that even the most advanced and the apparently most secure theory is not safe, that it can be modified or entirely overthrown with the help of views which the conceit of ignorance has already put into the dustbin of history. This is how the knowledge of today may become the fairy-tale of tomorrow and how the most laughable myth may eventually turn into the most solid piece of science.

Pluralism of theories and metaphysical views is not only important for methodology, it is also an essential part of a humanitarian outlook. Progressive educators have always tried to develop the individuality of their pupils and to bring to fruition the particular, and sometimes quite unique, talents and beliefs that a child possesses. Such an education, however, has very often seemed to be a futile exercise in day-dreaming. For is it not necessary to prepare the young for life *as it actually is*? Does this not mean that they must learn *one particular set of views* to the exclusion of everything else? And, if a trace of their imagination is still to remain, will it not find its proper application in the arts or in a thin domain of dreams that has but little to do with the world we live in? Will this procedure not finally lead to a split between a hated reality and welcome fantasies, science and the arts, careful description and unrestrained self-expression? The argument for proliferation shows that this need not happen. It is possible to *retain* what one might call the freedom of artistic creation *and to use it to the full*, not just as a road of escape but as a necessary means for discovering and perhaps even changing the features of the world we live in. This coincidence of the part (individual man) with the whole (the world we live in), of the purely subjective and arbitrary with the objective and lawful, is one of the most important arguments in favour of a pluralistic methodology. For details the reader is advised to consult Mill's magnificent essay *On Liberty*.[14]

14. Cf. my account of this essay in section 3 of 'Against Method', *Minnesota Studies in the Philosophy of Science*, Vol. 4, Minneapolis, 1970.

5

No theory ever agrees with all the facts in its domain, yet it is not always the theory that is to blame. Facts are constituted by older ideologies, and a clash between facts and theories may be proof of progress. It is also a first step in our attempt to find the principles implicit in familiar observational notions.

Considering now the invention, elaboration and use of theories which are inconsistent, not just with other theories, but even with *experiments, facts, observations*, we may start by pointing out that *no single theory ever agrees with all the known facts in its domain*. And the trouble is not created by rumours, or by the result of sloppy procedure. It is created by experiments and measurements of the highest precision and reliability.

It will be convenient, at this place, to distinguish two different kinds of disagreement between theory and fact: numerical disagreement, and qualitative failures.

The first case is quite familiar: a theory makes a certain numerical prediction and the value that is actually obtained differs from the prediction made by more than the margin of error. Precision instruments are usually involved here. Numerical disagreements abound in science. They give rise to an 'ocean of anomalies' that surrounds every single theory.[1]

Thus the Copernican view at the time of Galileo was inconsistent with facts so plain and obvious that Galileo had to call it 'surely false'.[2] 'There is no limit to my astonishment,' he writes in a later work,[3] 'when I reflect

1. For the 'ocean' and various ways of dealing with it, cf. my 'Reply to Criticism', *Boston Studies*, Vol. 2, 1965, pp. 224ff.

2. Galileo Galilei, *The Assayer*, quoted in *The Controversy on the Comets of 1618*, ed. S. Drake and C. D. O'Malley, London, 1960, p. 323.

3. Galileo Galilei, *Dialogue Concerning the Two Chief World Systems*, Berkeley, 1953, p. 328.

that Aristarchus and Copernicus were able to make reason so conquer sense that, in defiance of the latter, the former became mistress of their belief.' Newton's theory of gravitation was beset, from the very beginning, by difficulties serious enough to provide material for refutation. Even today and in the non-relativistic domain there 'exist numerous discrepancies between observation and theory'.[4] Bohr's atomic model was introduced, and retained, in the face of precise and unshakable contrary evidence.[5] The special theory of relativity was retained despite Kaufmann's unambiguous experimental results of 1906, and despite D. C. Miller's refutation (I am speaking of a refutation because the experiment was, from the point of view of contemporary evidence, at least as well performed as were the earlier experiments of Michelson and Morley).[6] The general theory of relativity, though surprisingly successful

4. Brower-Clemence, *Methods of Celestial Mechanics*, New York, 1961. Also R. H. Dicke, 'Remarks on the Observational Basis of General Relativity', *Gravitation and Relativity*, ed. H. Y. Chiu and W. F. Hoffman, New York, 1964, pp. 1–16. For a more detailed discussion of some of the difficulties of classical celestial mechanics, cf. J. Chazy, *La Théorie de la relativité et la Méchanique céleste*, Vol. I, Chapters 4 and 5, Paris, 1928.

5. Cf. Max Jammer, *The Conceptual Development of Quantum Mechanics*, New York, 1966, section 22. For an analysis cf. section 3c/2 of Lakatos, 'Falsification and the Methodology of Scientific Research Programmes', *Criticism and the Growth of Knowledge*, ed. Lakatos–Musgrave, Cambridge, 1970.

6. W. Kaufmann, 'Über die Konstitution des Elektrons', *Ann. Phys.*, No. 19, 1906, p. 487. Kaufmann stated his conclusion quite unambiguously, and in italics: '*The results of the measurements are not compatible with the fundamental assumption of Lorentz and Einstein.*' Lorentz' reaction: '. . . it seems very likely that we shall have to relinquish this idea altogether' (*Theory of Electrons*, second edition, p. 213). Ehrenfest: 'Kaufmann demonstrates that Lorentz' deformable electron is ruled out by the measurements' ('Zur Stabilitätsfrage bei den Bucherer–Langevin Elektronen', *Phys. Zs.*, Vol. 7, 1906, p. 302). Poincaré's reluctance to accept the 'new mechanics' of Lorentz can be explained at least in part, by the outcome of Kaufmann's experiment. Cf. *Science and Method*, New York, 1960, Book III, Chapter 2, section v, where Kaufmann's experiment is discussed in detail, the conclusion being that the 'principle of relativity . . . cannot have the fundamental importance one was inclined to ascribe to it'. Cf. also St. Goldberg, 'Poincaré's Silence and Einstein's Relativity', *British Journal for the History of Science*, Vol. 5, 1970, pp. 73ff, and the literature given there. Einstein alone regarded the results as 'improbable because their basic assumption, from which the mass of the moving electron is deduced, are not suggested by theoretical systems which encompass wider complexes of phenomena' (*Jahrbuch der Radioaktivität und Elektrizität*, Vol. 4, 1907, p. 439). Miller's work was studied by Lorentz for many years, but he could not find the trouble. It was only in 1955, twenty-five years after Miller had finished his experiments,

in some domains (see, however, the remarks below), failed to explain 10″ in the movement of the nodes of Venus and more than 5″ in the movement of the nodes of Mars;[7] moreover, it is now again in trouble, due to the new calculations on the motion of Mercury by Dicke and others.[8] All those are quantitative difficulties, which can be resolved by discovering a better set of *numbers* but which do not force us to make qualitative adjustments.[9]

---

that a satisfactory account of Miller's results was found. Cf. R. S. Shankland, 'Conversations with Einstein', *Am. Journ. Phys.*, Vol. 31, 1963, pp. 47–57, especially p. 51, as well as footnotes 19 and 34; cf. also the inconclusive discussion at the 'Conference on the Michelson–Morley Experiment', *Astrophysical Journal*, Vol. 68, 1928, pp. 341ff.

7. J. Chazy, op. cit., p. 230.

8. See R. H. Dicke, op. cit. Note that the later corrections of Dicke do not impair the argument which is that superseded theories (such as classical celestial mechanics) can be used for the criticism of their most successful replacements (general relativity). Besides, Dicke was a *temporary* danger, and that is all we need to know.

9. Herbert Feigl (*Minnesota Studies*, 5, 1971, p. 7) and Karl Popper (*Objective Knowledge*, p. 78) have tried to turn Einstein into a naive falsificationist. Thus Feigl writes: 'If Einstein relied on "beauty", "harmony", "symmetry", "elegance" in constructing . . . his general theory of relativity, it must nevertheless be remembered that he also said (in a lecture in Prague in 1920 – I was present then as a very young student): "If the observations of the red shift in the spectra of massive stars don't come out quantitatively in accordance with the principles of general relativity, then my theory will be dust and ashes".' Popper writes: 'Einstein . . . said that if the red shift effect . . . was not observed in the case of white dwarfs, his theory of general relativity would be refuted.'

Popper gives no source for his story, and he most likely has it from Feigl. But Feigl's story and Popper's repetition conflict with the numerous occasions where Einstein emphasizes the 'reason of the matter' ('die Vernunft der Sache') over and above 'verification by little effects' and this not only in casual remarks during a lecture, but in writing. Cf. the quotation in footnote 6 above, which deals with difficulties of the special theory of relativity and precedes the talk at which Feigl was present. Cf. also the letters to M. Besso and K. Seelig as quoted in G. Holton, 'Influences on Einstein's Early Work', *Organon*, No. 3, 1966, p. 242, and K. Seelig, *Albert Einstein*, Zürich, 1960, p. 271. In 1952 Born writes to Einstein as follows (*Born–Einstein Letters*, New York, 1971, p. 190, dealing with Freundlich's analysis of the bending of light near the sun and the red shift): 'It really looks as if your formula is not quite correct. It looks even worse in the case of the red shift [the crucial case referred to by Feigl and Popper]; this is much smaller than the theoretical value towards the centre of the sun's disk, and much larger at the edges. . . . Could this be a hint of non-linearity?' Einstein (letter of 12 May 1952, op. cit., p. 192) replies: 'Freundlich . . . does not move me in the slightest. Even if the deflection of light, the perihelial movement or line shift were

The second case, the case of qualitative failures, is less familiar, but of much greater interest. In this case a theory is inconsistent not with a recondite fact, that can be unearthed with the help of complex equipment and is known to experts only, but with circumstances which are easily noticed and which are familiar to everyone.

The first and, to my mind, the most important example of an inconsistency of this kind is Parmenides' theory of the unchanging and homogeneous One which is contradicted by almost everything we know and experience. The theory has much in its favour[10] and plays a role even today, for example in the general theory of relativity. Used in an undeveloped form by Anaximander, it led to the insight repeated by Heisenberg[11] in his theory of elementary particles that the basic substance, or the basic elements of the universe, cannot obey the same laws as the visible elements. The theory was supported by Zeno's arguments, which showed the difficulties inherent in the idea of a continuum consisting of isolated elements. Aristotle took these arguments seriously and developed his own theory of the continuum.[12] Yet the concept of continuum as a collection of elements remained and continued to be used,

---

unknown, the gravitation equations would still be convincing because they avoid the inertial system (the phantom which affects everything but is not itself affected). *It is really strange that human beings are normally deaf to the strongest arguments while they are always inclined to overestimate measuring accuracies*' (my italics). How is this conflict (between Feigl's testimony and Einstein's writings) to be explained? It cannot be explained by a *change* in Einstein's attitude. His disrespectful attitude towards observation and experiment was there from the very beginning, as we have seen. It might be explained either by a mistake on Feigl's part, or else as another instance of Einstein's 'opportunism' – cf. text to footnote 6 of the *Introduction*.

10. For a defence of Parmenidean procedures cf. my 'In Defence of Classical Physics', loc. cit. Cf. also the section on Parmenides in my *Einführung in die Naturphilosophie*.

11. W. Heisenberg, 'Der gegenwärtige Stand der Theorie der Elementarteilchen', *Naturwissenschaften*, No. 42, 1955, pp. 640ff. For a comprehensive account of Heisenberg's philosophy, cf. Herbert Hörz, *Werner Heisenberg und die Philosophie*, Berlin, 1966.

12. *Physics*, Book VI, *De Coelo*, 303a3ff; *De Generatione et Corruptione*, 316a. Aristotle's theory of the continuum seems to be closely connected with his empiricism. Aristotle's 'empiricism', however, is not just a philosophical dogma; it is a cosmological hypothesis that is clearly formulated (one hears, for a change, what kind of process experience is supposed to be) and leads, among other things, to a solution of problems which arose in other, and more 'metaphysical' traditions. The problem of the continuum seems to be one of those problems. For a survey of opinions on Zeno's paradoxes cf. *Zeno's Paradoxes*, ed. Salmon, New York, 1970.

despite the quite obvious difficulties, until these difficulties were almost removed early in the 20th century.[13]

A further example of a theory with qualitative defects is Newton's theory of colours. According to this theory, light consists of rays of different refrangibility which can be separated, reunited, refracted, but which are never changed in their internal constitution, and which have a very small lateral extension in space. Considering that the surface of mirrors is much rougher than the lateral extension of the rays, the ray theory is found to be inconsistent with the existence of mirror images (as is admitted by Newton himself): if light consists of rays, then a mirror should behave like a rough surface, i.e. it should look to us like a wall. Newton retained his theory, eliminating the difficulty with the help of an *ad hoc* hypothesis: 'the reflection of a ray is effected, not by a single point of the reflecting body, but by some power of the body which is evenly diffused all over its surface'.[14]

In Newton's case the qualitative discrepancy between theory and fact was removed by an *ad hoc* hypothesis. In other cases not even this very flimsy manoeuvre is used: one retains the theory *and tries to forget* its shortcomings. An example of this is the attitude towards Kepler's rule according to which an object viewed through a lens is perceived at the point at which the rays travelling from the lens towards the eye intersect.[15]

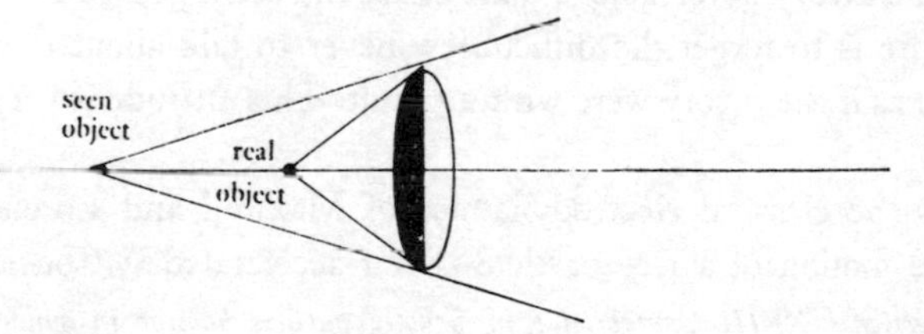

13. A. Grünbaum, 'A Consistent Conception of the Extended Linear Continuum as an Aggregate of Unextended Elements', *Philosophy of Science*, No. 19, 1952, p. 283, as well as the articles in Salmon, op. cit.

14. Sir Isaac Newton, *Optics*, Book 2, part 3, proposition 8, New York, 1952, p. 266. For a discussion of this aspect of Newton's method cf. my essay, 'Classical Empiricism', op. cit.

15. Johannes Kepler, *Ad Vitellionem Paralipomena, Johannes Kepler, Gesammelte Werke*, Vol. 2, München, 1939, p. 72. For a detailed discussion of Kepler's rule and its influence see Vasco Ronchi, *Optics: The Science of Vision*, New York, 1957, Chapters 43ff. Cf. also Chapters 9–11 below.

The rule implies that an object situated at the focus will be seen infinitely far away.

'But on the contrary,' writes Barrow, Newton's teacher and predecessor at Cambridge, commenting on this prediction,[16] 'we are assured by experience that [a point situated close to the focus] appears variously distant, according to the different situations of the eye.... And it does almost never seem further off than it would be if it were beheld with the naked eye; but, on the contrary, it does sometimes appear much nearer.... All which does seem repugnant to our principles.' 'But for me,' Barrow continues, 'neither this nor any other difficulty shall have so great an influence on me, as to make me renounce that which I know to be manifestly agreeable to reason.'

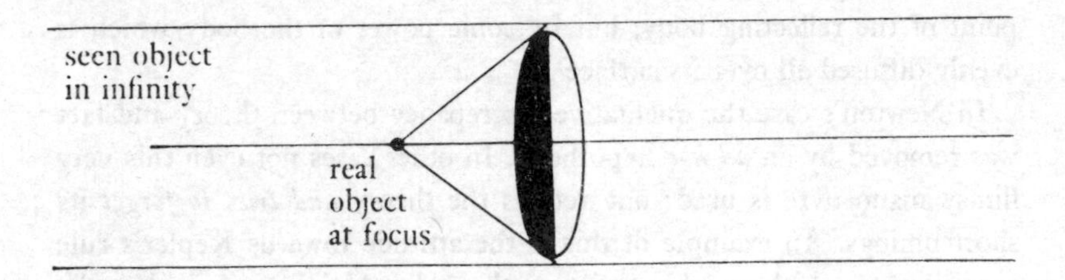

Barrow *mentions* the qualitative difficulties, and he *says* that he will retain the theory nevertheless. This is not the usual procedure. The usual procedure is to forget the difficulties, never to talk about them, and to proceed as if the theory were without fault. This attitude is very common today.

Thus the classical electrodynamics of Maxwell and Lorentz implies that the motion of a free particle is self-accelerated.[17] Considering the

16. *Lectiones XVIII Cantabrigiae in Scholio publicis habitae in quibus Opticorum Phenomenon genuinae Rationes investigantur ac exponentur*, London, 1669, p. 125. The passage is used by Berkeley in his attack on the traditional, 'objectivistic' optics (*An Essay Towards a New Theory of Vision*, Works, Vol. 1, ed. Frazer, London, 1901, pp. 137ff).

17. Assuming M to be the observed mass of the charged particle we obtain for its acceleration at time t the value

$$b(t) = b(o) \,.\, \exp\left[\,{}^{3}\!/_{2}\ \frac{Mc^{3}}{e^{2}}\right] t.$$

Cf. D. H. Sen, *Fields and/or Particles*, New York, 1968, p. 10. For this particular difficulty, cf. also H. R. Post, 'Correspondence, Invariance and Heuristics' in *Studies in* self-energy of the electron one obtains divergent expressions for point-charges while charges of finite extension can be made to agree with relativity only by adding untestable stresses and pressures inside the electron.[18] The problem reappears in the quantum theory, though it is here partially covered up by 'renormalization'. This procedure consists in crossing out the results of certain calculations and replacing them by a description of what is actually observed. Thus one admits, implicitly, that the theory is in trouble while formulating it in a manner suggesting that a new principle has been discovered.[19] Small wonder when philosophically unsophisticated authors get the impression that 'all evidence points with merciless definiteness in the . . . direction . . . [that] all the processes involving . . . unknown interactions conform to the fundamental quantum law.'[20]

Another example of modern physics is quite instructive, for it might have led to an entirely different development of our knowledge concerning

---

the History and Philosophy of Science, November, 1971, footnote 14. Post's assertion that physics 'is remarkably unsuccessful . . .' as a science, p. 219, and that, by comparison, 'botany has fair predictive power in its own field of application', footnote 14, agrees with my own opinion and indicates that Aristotelian science, taken as a whole, may have been more adequate than its highly abstract successors. Post and I disagree on many other points, however. The reader is urged to read his brilliant essay as a partial antidote against the point of view which I try to defend.

18. Cf. W. Heitler, *The Quantum Theory of Radiation*, Oxford, 1954, p. 31.

19. Apart from this *methodological* objection, there are also *factual* difficulties. Cf. the discussion at the 12th Solvay Conference, *The Quantum Theory of Fields*, New York, 1962, especially the contributions by Heitler and Feynman. Today (1971) the situation is essentially the same, cf. Brodsky and Drell, 'The Present Status of Quantum Electrodynamics', *Annual Review of Nuclear Science*, Vol. 20, Palo Alto, 1970, p. 190. Each of the examples of footnotes 3–16 can be used as a basis for case studies of the kind to be carried out in Chapters 6–12 (Galileo and the Copernican Revolution). This shows that the case of Galileo is not 'an *exception* characterizing the beginning of the so-called scientific revolution' (G. Radnitzky, 'Theorienpluralismus Theorienmonismus' in *Der Methoden- und Theorienpluralismus in den Wissenschaften*, ed. Diemer Meisenheim, 1971, p. 164) but is *typical* of scientific change at all times. However, I agree with Radnitzky that 'today', that is in the physics of 1960/1970, the situation may be somewhat different. The reason is that physics is now experiencing a period of *stagnation*, where a tremendous increase in bulk covers up an astounding poverty in new fundamental ideas. (This stagnation is connected with the fact that physics is changing from a science into a business and that younger physicists no longer use history and philosophy as an instrument of research.)

20. Rosenfeld in *Observation and Interpretation*, London, 1957, p. 44.

the microcosm. Ehrenfest has proved a theorem according to which the classical electron theory of Lorentz taken in conjunction with the equipartition principle excludes induced magnetism.[21] The reasoning is exceedingly simple; according to the equipartition principle, the probability of a given motion is proportional to exp $(-U/RT)$, where U is the energy of the motion. Now the energy of an electron moving in a constant magnetic field B is, according to Lorentz, $U = Q(E + V \times B)$. V, where Q is the charge of the moving particle, V its velocity and E the electric field. This magnitude reduces to QEV in all cases unless one is prepared to admit the existence of single magnetic poles. (Given the proper context, this result strongly supports the ideas and experimental findings of the late Felix Ehrenhaft.[22])

Occasionally it is impossible to survey all the interesting consequences, and thus to discover the absurd results of a theory. This may be due to a deficiency in the existing mathematical methods; it may also be due to the ignorance of those who defend the theory. Under such circumstances, the most common procedure is to use an older theory up to a certain point (which is often quite arbitrary) and to add the new theory for calculating refinements. Seen from a methodological point of view the procedure is a veritable nightmare. Let us explain it using the relativistic calculation of the path of Mercury as an example.

The perihelion of Mercury moves along at a rate of about 5600″ per century. Of this value, 5026″ are geometric, having to do with the movement of the reference system, while 531″ are dynamical, due to perturbations in the solar system. Of these perturbations all but the famous 43″ are accounted for by classical mechanics. This is how the situation is usually explained.

The explanation shows that the premise from which we derive the 43″ is not the general theory of relativity plus suitable initial conditions.

21. This difficulty was realized by Bohr in his doctoral thesis, cf. Niels Bohr, *Collected Works*, Vol. I, Amsterdam, 1972, pp. 158, 381. He pointed out that the velocity changes due to the changes in the external field would equalize after the field was established, so that no magnetic effects could arise. Cf. also Heilbron and T. S. Kuhn, 'The Genesis of the Bohr Atom', *Historical Studies in the Physical Sciences*, No. 1, 1969, p. 221. The argument in the text is taken from *The Feynman Lectures*, Vol. 2, California and London, 1965, Chapter 34.6. For a somewhat clearer account cf. R. Becker, *Theorie der Elektrizität*, Leipzig, 1949, p. 132.

22. See footnote 5 to Chapter 3.

The premise contains classical physics *in addition* to whatever relativistic assumptions are being made. Furthermore, the relativistic calculation, the so-called 'Schwarzschild solution', does not deal with the planetary system as it exists in the real world (i.e. our own asymmetric galaxy); it deals with the entirely fictional case of a central symmetrical universe containing a singularity in the middle and nothing else. What are the reasons for employing such an odd conjunction of premises?

The reason, according to the customary reply, is that we are dealing with approximations. The formulae of classical physics do not appear because relativity is incomplete. Nor is the centrally symmetric case used because relativity does not offer anything better. Both schemata flow from the general theory under the special circumstances realized in our planetary system *provided* we omit magnitudes too small to be considered. Hence, we are using the theory of relativity throughout, and we are using it in an adequate manner.

Note how this idea of an approximation differs from the legitimate idea. Usually one has a theory, one is able to calculate the particular case one is interested in, one notes that this calculation leads to magnitudes below experimental precision, one omits such magnitudes, and one obtains a vastly simplified formalism. In the present case, making the required approximations would mean calculating the full *n*-body problem relativistically (including long-term resonances between different planetary orbits), omitting magnitudes smaller than the precision of observation reached, and showing that the theory thus curtailed coincides with classical celestial mechanics as corrected by Schwarzschild. This procedure has not been used by anyone simply because the relativistic *n*-body problem has as yet withstood solution. There are not even approximate solutions for important problems such as, for example, the problem of stability (one of the first great stumbling blocks for Newton's theory). The classical part of the explanans, therefore, does not occur just for convenience, *it is absolutely necessary*. And the approximations made are not a result of relativistic calculations, they are introduced in order to make relativity fit the case. One may properly call them *ad hoc approximations*.

Ad hoc approximations abound in modern mathematical physics. They play a very important part in the quantum theory of fields and they are an essential ingredient of the correspondence principle. At the moment we are not concerned with the reasons for this fact, we are only concerned with its consequences: *ad hoc* approximations conceal, and even entirely eliminate, qualitative difficulties. They create a false impression of the excellence of our science. It follows that a philosopher who wants to study the adequacy of science as a picture of the world, or who wants to build up a realistic scientific methodology, must look at modern science with special care. In most cases modern science is more opaque, and much more deceptive, than its 16th- and 17th-century ancestors have ever been.

As a final example of qualitative difficulties I mention again the heliocentric theory at the time of Galileo. I shall soon have occasion to show that this theory was inadequate both qualitatively and quantitatively, and that it was also philosophically absurd.

To sum up this brief and very incomplete list: wherever we look, whenever we have a little patience and select our evidence in an unprejudiced manner, we find that theories fail adequately to reproduce certain *quantitative results*, and that they are *qualitatively incompetent* to a surprising degree. Science gives us theories of great beauty and sophistication. Modern science has developed mathematical structures which exceed anything that has existed so far in coherence and generality. But in order to achieve this miracle all the existing troubles had to be pushed into the *relation* between theory and fact,[23] and had to be concealed, by *ad hoc* approximations and by other procedures.

23. Von Neumann's work in quantum mechanics is an especially instructive example of this procedure. In order to arrive at a satisfactory proof of the expansion theorem in Hilbert Space, von Neumann replaces the quasi-intuitive notions of Dirac (and Bohr) by more complex notions of his own. The theoretical relations between these notions are accessible to a more rigorous treatment than the theoretical relations between the notions that preceded them ('more rigorous' from the point of view of von Neumann and his followers). It is different with their relation to experimental procedures. No measuring instruments can be specified for the great majority of observables (Wigner, *American Journal of Physics*, Vol. 31, 1963, p. 14), and where specification is possible it becomes necessary to modify well known and unrefuted laws in an arbitrary way or else to admit that some quite ordinary problems of quantum mechanics, such as the scattering problem, do not have a solution (J. M. Cook, *Journal of Mathematical Physics*, Vol. 36, 1957). Thus the theory becomes a veritable monster of rigour and precision while its relation to experience is more obscure than ever. It is interesting to see that similar developments also occur in 'primitive thought'. 'The most striking

This being the case, what shall we make of the methodological demand that a theory must be judged by experience and must be rejected if it contradicts accepted basic statements? What attitude shall we adopt towards the various theories of confirmation and corroboration, which all rest on the assumption that theories can be made to agree completely with the known facts, and which use the amount of agreement reached as a principle of evaluation? This demand, these theories, are now all seen to be quite useless. They are as useless as a medicine that heals a patient only if he is bacteria-free. In practice they are never obeyed by anyone. Methodologists may point to the importance of falsifications – but they blithely use falsified theories; they may sermonize how important it is to consider all the relevant evidence, and never mention those big and drastic facts which show that the theories which they admire and accept, like the theory of relativity or the quantum theory, may be as badly off as the older theories which they reject. In *practice* they slavishly repeat the most recent pronouncements of the top dogs in physics, though in doing so they must violate some very basic rules of their trade. Is it possible to proceed in a more reasonable manner? Let us see![24]

According to Hume, theories cannot be *derived from* facts. The demand to admit only those theories which follow from facts leaves us without any theory. Hence, science *as we know it* can exist only if we drop the demand and revise our methodology.

According to our present results, hardly any theory is *consistent with* the facts. The demand to admit only those theories which are consistent with the available and accepted facts again leaves us without any theory. (I repeat: *without any theory*, for there is not a single theory that is not in some trouble or other.) Hence, a science as we know it can exist only if we drop this demand also and again revise our methodology, *now admitting*

---

features of Nupe sand divining,' writes S. F. Nader in *Nupe Religion*, 1954, p. 63, 'is the contrast between its pretentious theoretical framework and its primitive and slipshod application in practice.' It does not need a science to produce Neumannian nightmares.

24 The existence of qualitative difficulties, or 'pockets of resistance' (St Augustine, *Contra Julianum*, V, xiv, 51 – *Migne*, Vol. 44), was used by the Church fathers to defuse objections which the science of their time raised against parts of the Christian faith, such as the resurrection of Christ (which Porphyry had regarded as being incompatible with physics).

counterinduction in addition to admitting unsupported hypotheses. The right method must not contain any rules that make us choose between theories *on the basis of falsification.* Rather, its rules must enable us to choose between theories which we have already tested *and which are falsified.*

To proceed further. Not only are facts and theories in constant disharmony, they are never as neatly separated as everyone makes them out to be. Methodological rules speak of 'theories', 'observations' and 'experimental results' as if these were clear-cut well-defined objects whose properties are easy to evaluate and which are understood in the same way by all scientists.

However, the material which a scientist *actually* has at his disposal, his laws, his experimental results, his mathematical techniques, his epistemological prejudices, his attitude towards the absurd consequences of the theories which he accepts, is indeterminate in many ways, ambiguous, *and never fully separated from the historical background.* This material is always contaminated by principles which he does not know and which, if known, would be extremely hard to test. Questionable views on cognition such as the view that our senses, used in normal circumstances, give reliable information about the world, may invade the observation language itself, constituting the observational terms as well as the distinction between veridical and illusory appearance. As a result, observation languages may become tied to older layers of speculation which affect, in this roundabout fashion, even the most progressive methodology. (Example: the absolute space–time frame of classical physics which was codified and consecrated by Kant.) The sensory impression, however simple, always contains a component that expresses the physiological reaction of the perceiving organism and has no objective correlate. This 'subjective' component often merges with the rest, and forms an unstructured whole which must then be subdivided from the outside with the help of counterinductive procedures. (An example of this is the appearance of a fixed star to the naked eye, which contains the subjective effects of irradiation, diffraction, diffusion, restricted by the lateral inhibition of adjacent elements of the retina.) Finally, there are the auxiliary premises which are needed for the derivation of testable conclusions, and which occasionally form entire *auxiliary sciences.*

Consider the case of the Copernican hypothesis, whose invention, defence, and partial vindication runs counter to almost every methodological rule one might care to think of today. The auxiliary sciences here contained laws describing the properties and the influence of the terrestrial atmosphere (meteorology); optical laws dealing with the structure of the eye and of telescopes, and with the behaviour of light; and dynamical laws describing motion in moving systems. Most importantly, however, the auxiliary sciences contained a theory of cognition that postulated a certain simple relation between perceptions and physical objects. Not all these auxiliary disciplines were available in explicit form. Many of them merged with the observation language, and led to the situation described at the beginning of the preceding paragraph.

Consideration of all these circumstances, of observation terms, sensory core, auxiliary sciences, background speculation, suggest that a theory may be inconsistent with the evidence, not because it is incorrect, *but because the evidence is contaminated.* The theory is threatened because the evidence either contains unanalysed sensations which only partly correspond to external processes, or because it is presented in terms of antiquated views, or because it is evaluated with the help of backward auxiliary subjects. The Copernican theory was in trouble for *all* these reasons.

It is this *historico-physiological character of the evidence*, the fact that it does not merely describe some objective state of affairs *but also expresses some subjective, mythical, and long-forgotten views* concerning this state of affairs, that forces us to take a fresh look at methodology. It shows that it would be extremely imprudent to let the evidence judge our theories directly and without any further ado. A straightforward and unqualified judgement of theories by 'facts' is bound to eliminate ideas *simply because they do not fit into the framework of some older cosmology*. Taking experimental results and observations for granted and putting the burden of proof on the theory means taking the observational ideology for granted without having ever examined it. (Note that the experimental results are supposed to have been obtained with the greatest possible care. Hence 'taking observations, etc., for granted' means 'taking them for granted *after* the most careful examination of their reliability': for even the most careful examination of an observation statement does not interfere with the concepts in which it is expressed, or with the structure of the sensory image.)

Now – how can we possibly examine something we use all the time and presuppose in every statement? How can we criticise the terms in which we habitually express our observations? Let us see!

The first step in our criticism of commonly-used concepts is to create a measure of criticism, something with which these concepts can be *compared*. Of course, we shall later want to know a little more about the measuring-stick itself; for example, we shall want to know whether it is better than, or perhaps not as good as, the material examined. But in order for *this* examination to start there must be a measuring-stick in the first place. Therefore, the first step in our criticism of customary concepts and customary reactions is to step outside the circle and either to invent a new conceptual system, for example a new theory, that clashes with the most carefully established observational results and confounds the most plausible theoretical principles, or to import such a system from outside science, from religion, from mythology, from the ideas of incompetents,[25] or the ramblings of madmen. This step is, again, counterinductive. Counterinduction is thus both a *fact* – science could not exist without it – and a legitimate and much needed *move* in the game of science.

25. It is interesting to see that Philolaos who disregarded the evidence of the senses and set the earth in motion was 'an unmathematical confusionist. It was the confusionist who found the courage lacking in many great observers and mathematically well-informed scientists to disregard the immediate evidence of the senses in order to remain in agreement with principles he firmly believed.' K. von Fritz, *Grundprobleme der Geschichte der antiken Wissenschaft*, Berlin–New York, 1971, p. 165. 'It is therefore not surprising that the next step on this path was due to a man whose writings, as far as we know them, show him as a talented stylist and popularizer with occasionally interesting ideas of his own rather than as a profound thinker or exact scientist,' op cit., p. 184. Confusionists and superficial intellectuals *move ahead* while the 'deep' thinkers *descend* into the darker regions of the status quo or, to express it in a different way, they remain stuck in the mud.

6

As an example of such an attempt I examine the tower argument *which the Aristotelians used to refute the motion of the earth. The argument involves* natural interpretations – *ideas so closely connected with observations that it needs a special effort to realize their existence and to determine their content. Galileo identifies the natural interpretations which are inconsistent with Copernicus and replaces them by others.*

It seems to me that (Galileo) suffers greatly from continual digressions, and that he does not stop to explain all that is relevant at each point; which shows that he has not examined them in order, and that he has merely sought reasons for particular effects, without having considered . . . first causes . . .; and thus that he has built without a foundation.

DESCARTES

I am (indeed) unwilling to compress philosophical doctrines into the most narrow kind of space and to adopt that stiff, concise and graceless manner, that manner bare of any adornment which pure geometricians call their own, not uttering a single word that has not been given to them by strict necessity . . . I do not regard it as a fault to talk about many and diverse things, even in those treatises which have only a single topic . . . for I believe that what gives grandeur, nobility, and excellence to our deeds and inventions does not lie in what is necessary – though the absence of it would be a great mistake – but in what is not. . . .

GALILEO

But where common sense believes that rationalizing sophists have the intention of shaking the very fundament of the commonweal,

then it would seem to be not only reasonable, but permissible, and
even praiseworthy to aid the good cause with sham
reasons rather than leaving the advantage to the . . . opponent.

KANT[1]

As a concrete illustration and as a basis for further discussion, I shall now briefly describe the manner in which Galileo defused an important counter-argument against the idea of the motion of the earth. I say 'defused', and not 'refuted', because we are dealing with a changing conceptual system as well as with certain attempts at concealment.

According to the argument which convinced Tycho, and which is used against the motion of the earth in Galileo's own *Trattato della sfera*, observation shows that 'heavy bodies . . . falling down from on high, go by a straight and vertical line to the surface of the earth. This is considered an irrefutable argument for the earth being motionless. For, if it made the diurnal rotation, a tower from whose top a rock was let fall, being carried by the whirling of the earth, would travel many hundreds of yards to the east in the time the rock would consume in its fall, and

1. The three quotations are: Descartes, letter to Mersenne of 11 October 1638, *Oeuvres*, II, p. 380. Galileo, letter to Leopold of Toscana of 1640, usually quoted under the title *Sul Candor Lunare*, *Edizione Nazionale*, VIII, p. 491. For a detailed discussion of Galileo's style and its connection with his natural philosophy cf. L. Olschki, *Galileo und seine Zeit: Geschichte der neusprachlichen wissenschaftlichen Literatur*, Vol. III, Halle, 1927, reprinted Vaduz, 1965. The letter to Leopold is quoted and discussed on pp. 455ff.

Descartes' letter is discussed by Salmon as an example of the issue between rationalism and empiricism in 'The Foundations of Scientific Inference', *Mind and Cosmos*, ed. Colodny, Pittsburgh, 1966, p. 136. It should rather be regarded as an example of the issue between dogmatic methodologies and opportunistic methodologies, bearing in mind that empiricism can be as strict and unyielding as the most rigorous types of rationalism.

The Kant quotation is from the *Critique of Pure Reason*, B 777, 8ff (the quotation was brought to my attention by Professor Stanley Rosen's work on Plato's *Symposion*). Kant continues: 'However, I would think that there is nothing that goes less well together with the intention of asserting a good cause than subterfuge, conceit, and deception. *If* one could take only this much for granted, then the battle of speculative reason . . . would have been concluded long ago, or would soon come to an end. Thus the purity of a cause often stands in the inverse proportion to its truth. . . .' One should also note that Kant explains the rise of *civilization* on the basis of disingenuous moves which 'have the function to raise mankind above its crude past', op. cit., 776, 14f. Similar ideas occur in his account of world history.

the rock ought to strike the earth that distance away from the base of the tower.'[2]

In considering the argument, Galileo at once admits the correctness of the sensory content of the observation made, viz. that 'heavy bodies . . . falling from a height, go perpendicularly to the surface of the earth'.[3] Considering an author (Chiaramonti) who sets out to convert Copernicans by repeatedly mentioning this fact, he says: 'I wish that this author would not put himself to such trouble trying to have us understand from our senses that this motion of falling bodies is simple straight motion and no other kind, nor get angry and complain because such a clear, obvious, and manifest thing should be called into question. For in this way he hints at believing that to those who say such motion is not straight at all, but rather circular, it seems they see the stone move visibly in an arc, since he calls upon their senses rather than their reason to clarify the effect. This is not the case, Simplicio; for just as I . . . have never seen nor ever expect to see, the rock fall any way but perpendicularly, just so do I believe that it appears to the eyes of everyone else. It is, therefore, better to put aside the appearance, on which we all agree, and to use the power of reason either to confirm its reality or to reveal its fallacy.'[4] The correctness of the observation is not in question. What is in question is its 'reality' or 'fallacy'. What is meant by this expression?

The question is answered by an example that occurs in Galileo's next paragraph, 'from which . . . one may learn how easily anyone may be deceived by simple appearance, or let us say by the impressions of one's senses. This event is the appearance to those who travel along a street by night of being followed by the moon, with steps equal to theirs, when they see it go gliding along the eaves of the roofs. There it looks to them just as would a cat really running along the tiles and putting them behind it; an appearance which, if reason did not intervene, would only too obviously deceive the senses.'

In this example, we are asked to start with a sensory impression and to consider a statement that is forcefully suggested by it. (The suggestion is so strong that it has led to entire systems of belief and to rituals, as becomes clear from a closer study of the lunar aspects of witchcraft and of other cosmological hypotheses.) Now 'reason intervenes'; the

2. *Dialogue*, op. cit., p. 126. 3. ibid., p. 125. 4. ibid., p. 256.

statement suggested by the impression is examined, and one considers other statements in its place. The nature of the impression is not changed a bit by this activity. (This is only approximately true; but we can omit for our present purpose the complications arising from an interaction of impression and proposition.) But it enters new observation statements and plays new, better or worse, parts in our knowledge. What are the reasons and the methods which regulate such exchange?

To start with, we must become clear about the nature of the total phenomenon: appearance plus statement. There are not two acts – one, noticing a phenomenon; the other, expressing it with the help of the appropriate statement – *but only one*, viz. saying in a certain observational situation, 'the moon is following me', or, 'the stone is falling straight down'. We may, of course, abstractly subdivide this process into parts, and we may also try to create a situation where statement and phenomenon seem to be psychologically apart and waiting to be related. (This is rather difficult to achieve and is perhaps entirely impossible.) But under normal circumstances such a division does not occur; describing a familiar situation is, for the speaker, an event in which statement and phenomenon are firmly glued together.

This unity is the result of a process of learning that starts in one's childhood. From our very early days we learn to react to situations with the appropriate responses, linguistic or otherwise. The teaching procedures both *shape* the 'appearance', or 'phenomenon', and establish a firm *connection* with words, so that finally the phenomena seem to speak for themselves without outside help or extraneous knowledge. They *are* what the associated statements assert them to be. The language they 'speak' is, of course, influenced by the beliefs of earlier generations which have been held for so long that they no longer appear as separate principles, but enter the terms of everyday discourse, and, after the prescribed training, seem to emerge from the things themselves.

At this point we may want to compare, in our imagination and quite abstractly, the results of the teaching of different languages incorporating different ideologies. We may even want consciously to change some of these ideologies and adapt them to more 'modern' points of view. It is very difficult to say how this will alter our situation, *unless* we make the further assumption that the quality and structure of sensations (perceptions) or at least the quality and structure of those sensations which enter the body of science, is independent of their linguistic expression. I am very doubtful about even the approximate validity of this assumption, which can be refuted by simple examples, and I am sure that we are depriving ourselves of new and surprising discoveries as long as we remain within the limits defined by it. Yet, I shall for the moment, remain quite consciously within these limits. (My first task, if I should ever resume writing, would be to explore these limits and to venture beyond them.)

Making the additional simplifying assumption, we can now distinguish between sensations and those 'mental operations which follow so closely upon the senses',[5] and which are so firmly connected with their reactions that a separation is difficult to achieve. Considering the origin and the effect of such operations, I shall call them *natural interpretations*.

In the history of thought, natural interpretations have been regarded either as *a priori presuppositions* of science, or else as *prejudices* which must be removed before any serious examination can begin. The first view is that of Kant, and, in a very different manner and on the basis of very different talents, that of some contemporary linguistic philosophers. The second view is due to Bacon (who had predecessors, however, such as the Greek sceptics).

Galileo is one of those rare thinkers who neither wants forever to *retain* natural interpretations nor altogether to *eliminate* them. Wholesale judgements of this kind are quite alien to his way of thinking. He insists upon a *critical discussion* to decide which natural interpretations can be kept and which must be replaced. This is not always clear from his writings. Quite the contrary. The methods of reminiscence, to which he appeals so freely, are designed to create the impression that nothing has changed and that we continue expressing our observations in old and familiar ways. Yet his attitude is relatively easy to ascertain: natural interpretations are *necessary*. The senses alone, without the help of reason, cannot give us a true account of nature. What is needed for arriving at such a true account are 'the . . . senses, *accompanied by reasoning*'.[6]

5. Francis Bacon, *Novum Organum*, Introduction.
6. *Dialogue*, op. cit., p. 255, my italics.

Moreover, in the arguments dealing with the motion of the earth, it is this reasoning, it is the connotation of the observation terms and *not* the message of the senses or the appearance that causes trouble. 'It is, therefore, better to put aside the appearance, on which we all agree, and to use the power of reason either to confirm its reality or to reveal its fallacy.'[7] Confirming the reality or revealing the fallacy of appearances means, however, examining the validity of those natural interpretations which are so intimately connected with the appearances that we no longer regard them as separate assumptions. I now turn to the first natural interpretation implicit in the argument from falling stones.

According to Copernicus the motion of a falling stone should be 'mixed straight-and-circular'.[8] By the 'motion of the stone', is meant not just its motion relative to some visible mark in the visual field of the observer, or its observed motion, but rather its motion in the solar system or in (absolute) space, i.e. its *real motion*. The familiar facts appealed to in the argument assert a different kind of motion, a simple vertical motion. This result refutes the Copernican hypothesis only if the concept of motion that occurs in the observation statement is the same as the concept of motion that occurs in the Copernican prediction. The observation statement 'the stone is falling straight down' must, therefore, refer to a movement in (absolute) space. It must refer to a real motion.

Now, the force of an 'argument from observation' derives from the fact that the observation statements involved are firmly connected with appearances. There is no use appealing to observation if one does not know how to describe what one sees, or if one can offer one's description with hesitation only, as if one had just learned the language in which it is formulated. Producing an observation statement, then, consists of two very different psychological events: (1) a clear and unambiguous *sensation* and (2) a clear and unambiguous *connection* between this sensation and parts of a language. This is the way in which the sensation is made to speak. Do the sensations in the above argument speak the language of real motion?

They speak the language of real motion in the context of 17th-century everyday thought. At least, this is what Galileo tells us. He tells us that

7. ibid., p. 256. 8. ibid., p. 248.

the everyday thinking of the time assumes the 'operative' character of *all* motion, or, to use well-known philosophical terms, it assumes *a naive realism with respect to motion*: except for occasional and unavoidable illusions, apparent motion is identical with real (absolute) motion. Of course, this distinction is not explicitly drawn. One does not first distinguish the apparent motion from the real motion and then connect the two by a correspondence rule. One rather describes, perceives, acts towards motion as if it were already the real thing. Nor does one proceed in this manner under all circumstances. It is admitted that objects may move which are not seen to move; and it is also admitted that certain motions are illusory (cf. the example of the moon mentioned earlier in this chapter). Apparent motion and real motion are not always identified. However, there are *paradigmatic cases* in which it is psychologically very difficult, if not plainly impossible, to admit deception. It is from these paradigmatic cases, and not from the exceptions, that naive realism with respect to motion derives its strength. These are also the situations in which *we* first learn our kinematic vocabulary. From our very childhood we learn to react to them with concepts which have naive realism built right into them, and which inextricably connect movement and the appearance of movement. The motion of the stone in the tower argument, or the alleged motion of the earth, is such a paradigmatic case. How could one possibly be unaware of the swift motion of a large bulk of matter such as the earth is supposed to be! How could one possibly be unaware of the fact that the falling stone traces a vastly extended trajectory through space! From the point of view of 17th-century thought and language, the argument is, therefore, impeccable and quite forceful. However, notice how *theories* ('operative character' of all motion; essential correctness of sense reports) which are not formulated explicitly, enter the debate in the guise of observational terms. We realize again that observational terms are Trojan horses which must be watched most carefully. How is one supposed to proceed in such a sticky situation?

The argument from falling stones seems to refute the Copernican view. This may be due to an inherent disadvantage of Copernicanism; but it may also be due to the presence of natural interpretations which are in need of improvement. The first task, then, is to *discover* and to isolate these unexamined obstacles to progress.

It was Bacon's belief that natural interpretations could be discovered by a method of analysis that peels them off, one after another, until the sensory core of every observation is laid bare. This method has serious drawbacks. First, natural interpretations of the kind considered by Bacon are not just *added to* a previously existing field of sensations. They are instrumental in *constituting* the field, as Bacon says himself. Eliminate all natural interpretations, and you also eliminate the ability to think and to perceive. Second, disregarding this fundamental function of natural interpretations, it should be clear that a person who faces a perceptual field without a single natural interpretation at his disposal would be *completely disoriented*, he could not even *start* the business of science. The fact that we *do* start, even after some Baconian analysis, therefore shows that the analysis has stopped prematurely. It has stopped at precisely those natural interpretations of which we are not aware and without which we cannot proceed. It follows that the intention to start from scratch, after a complete removal of all natural interpretations, is self-defeating.

Furthermore, it is not possible even *partly* to unravel the cluster of natural interpretations. At first sight the task would seem to be simple enough. One takes observation statements, one after the other, and analyses their content. However, concepts that are hidden in observation statements are not likely to reveal themselves in the more abstract parts of language. If they do, it will still be difficult to nail them down; concepts, just like percepts, are ambiguous and dependent on background. Moreover, the content of a concept is determined also by the way in which it is related to perception. Yet, how can this way be discovered without circularity? Perceptions must be identified, and the identifying mechanism will contain some of the very same elements which govern the use of the concept to be investigated. We never penetrate this concept completely, for we always use part of it in the attempt to find its constituents. There is only one way to get out of this circle, and it consists in using an *external measure of comparison*, including new ways of relating concepts and percepts. Removed from the domain of natural discourse and from all those principles, habits, and attitudes which constitute its form of life, such an external measure will look strange indeed. This, however, is not an argument against its use. On the contrary, such an impression of strangeness reveals that natural interpretations are at work, and it is a first step towards their discovery. Let us explain this situation with the help of the tower example.

The example is intended to show that the Copernican view is not in accordance with 'the facts'. Seen from the point of view of these 'facts', the idea of the motion of the earth is outlandish, absurd, and obviously false, to mention only some of the expressions which were frequently used at the time, and which are still heard whenever professional squares confront a new and counter-factual theory. This makes us suspect that the Copernican view is an external measuring rod of precisely the kind described above.

We can now turn the argument around and use it as a *detecting device* that helps us to discover the natural interpretations which exclude the motion of the earth. Turning the argument around, we *first assert* the motion of the earth and *then inquire* what changes will remove the contradiction. Such an inquiry may take considerable time, and there is a good sense in which it is not finished even today. The contradiction, therefore, may stay with us for decades or even centuries. Still, *it must be upheld* until we have finished our examination or else the examination, the attempt to discover the antediluvian components of our knowledge, cannot even start. This, we have seen, is one of the reasons one can give for *retaining*, and, perhaps, even for *inventing*, theories which are inconsistent with the facts. Ideological ingredients of our knowledge and, more especially, of our observations, are discovered with the help of theories which are refuted by them. *They are discovered counterinductively*.

Let me repeat what has been asserted so far. Theories are tested, and possibly refuted, by facts. Facts contain ideological components, older views which have vanished from sight or were perhaps never formulated in an explicit manner. Such components are highly suspicious. Firstly, because of their age and obscure origin: we do not know why and how they were first introduced; secondly, because their very nature protects them, and always has protected them, from critical examination. In the event of a contradiction between a new and interesting theory and a collection of firmly established facts, the best procedure, therefore, is not to abandon the theory but to use it to discover the hidden principles responsible for the contradiction. Counterinduction is an essential part of such a process of discovery. (Excellent historical example: the arguments against motion and atomicity of Parmenides and Zeno. Diogenes of Sinope, the Cynic, took the simple course that would be taken by many contemporary scientists and all contemporary philosophers: he refuted the arguments by rising and walking up and down. The opposite course, recommended here, has led to much more interesting results, as is witnessed by the history of the case. One should not be too hard on Diogenes, however, for it is also reported that he beat up a pupil who was content with his refutation, exclaiming that he had given reasons which the pupil should not accept without additional reasons of his own.[9])

Having *discovered* a particular natural interpretation, how can we *examine* it and *test* it? Obviously, we cannot proceed in the usual way, i.e. derive predictions and compare them with 'results of observation'. These results are no longer available. The idea that the senses, employed under normal circumstances, produce correct reports of real events, for example reports of the real motion of physical bodies, has now been removed from all observational statements. (Remember that this notion was found to be an essential part of the anti-Copernican argument.) But without it our sensory reactions cease to be relevant for tests. This conclusion has been generalized by some older rationalists, who decided to build their science on reason only and ascribed to observation a quite insignificant auxiliary function. Galileo does not adopt this procedure.

If *one* natural interpretation causes trouble for an attractive view, and if its *elimination* removes the view from the domain of observation, then the only acceptable procedure is to use *other* interpretations and to see what happens. The interpretation which Galileo uses restores the senses to their position as instruments of exploration, *but only with respect to the reality of relative motion.* Motion 'among things which share it in common' is 'non-operative', that is, 'it remains insensible, imperceptible, and without any effect whatever'.[10] Galileo's first step, in his joint

9. Hegel, *Vorlesungen über die Geschichte der Philosophie*, I, ed. C. L. Michelet, Berlin, 1840, p. 289.

10. *Dialogue*, op. cit., p. 171. Galileo's kinematic relativism is not consistent. In the passage quoted, he proposes the view (1) that shared motion has *no effect whatsoever*. 'Motion,' he says, 'in so far as it is and acts as motion, to that extent exists relatively to things that lack it; and among things which all share equally in any motion, it does

examination of the Copernican doctrine and of a familiar but hidden natural interpretation, consists therefore in *replacing the latter by a different interpretation*. In other words, *he introduces a new observation language*.

This is, of course, an entirely legitimate move. In general, the observation language which enters an argument has been in use for a long time and is quite familiar. Considering the structure of common idioms on the one hand, and of the Aristotelian philosophy on the other, neither this use nor this familiarity can be regarded as a test of the underlying principles. These principles, these natural interpretations, occur in every description. Extraordinary cases which might create difficulties are defused with the help of 'adjustor words',[11] such as 'like' or 'analogous', which divert them so that the basic ontology remains unchallenged. A test is, however, urgently needed. It is especially needed in those cases where the principles seem to threaten a new theory. It is then quite reasonable to introduce alternative observation languages and to compare them both with the original idiom and with the theory under examination.

---

not act and is as if it did not exist' (p. 116); 'Whatever motion comes to be attributed to the earth must necessarily remain imperceptible . . . so long as we look only at terrestrial objects' (p. 114); '. . . motion that is common to many moving things is idle and inconsequential to the relation of those movables among themselves . . .' (p. 116). On the other hand, (2) he also suggests that 'nothing . . . *moves in a straight line by nature*. The motion of all celestial objects is in a circle; ships, coaches, horses, birds, all move in a circle around the earth; the motions of the parts of animals are all circular; in sum – we are forced to assume that only *gravia deorsum* and *levia sursum* move apparently in a straight line; but even that is not certain as long as it has not been proven that the earth is at rest' (p. 19). Now, if (2) is adopted, then the loose parts of systems moving in a straight line will tend to describe circular paths, thus contradicting (1). It is this inconsistency which has prompted me to split Galileo's argument into two steps, one dealing with the relativity of motion (only relative motion *is noticed*), the other dealing with inertial laws (and only inertial motion *leaves the relation between the parts of a system unaffected* – assuming, of course, that neighbouring inertial motions are approximately parallel). For the two steps of the argument, see the next chapter. One must also realize that accepting relativity of motion even for inertial paths, means giving up the *impetus theory*. This Galileo seems to have done by now, for his argument for the existence of 'boundless' or 'perpetual' motions which he outlines on pp. 147ff of the *Dialogue* appeals to motions which are neutral, i.e. neither natural nor forced, and which may therefore (?) be assumed to go on for ever.

11. J. L. Austin, *Sense and Sensibilia*, New York, 1964, p. 74. Adjustor words play an important role in the Aristotelian philosophy.

Proceeding in this way, we must make sure that the comparison is *fair*. That is, we must not criticize an idiom that is supposed to function as an observation language because it is not yet well known and is, therefore, less strongly connected with our sensory reactions and less plausible than is another, more 'common' idiom. Superficial criticisms of this kind, which have been elevated into an entire new 'philosophy' abound in discussions of the mind–body problem. Philosophers who want to introduce and to test new views thus find themselves faced not with *arguments*, which they could most likely answer, but with an impenetrable stone wall of well-entrenched *reactions*. This is not at all different from the attitude of people ignorant of foreign languages, who feel that a certain colour is much better described by 'red' than by 'rosso'. As opposed to such attempts at conversion by appeal to familiarity ('I *know* what pains are, and I also *know*, from introspection, that they have nothing whatever to do with material processes!'), we must emphasize that a comparative judgement of observation languages, e.g. materialistic observation languages, phenomenalistic observation languages, objective-idealistic observation languages, theological observation languages, etc., can start only *when all of them are spoken equally fluently*.

Let us now continue with our analysis of Galileo's reasoning.

7

The new natural interpretations constitute a new and highly abstract observation language. They are introduced and concealed *so that one fails to notice the change that has taken place (method of anamnesis). They contain the idea of the* relativity of all motion *and the* law of circular inertia.

Galileo replaces one natural interpretation by a very different and as yet (1630) at least partly unnatural interpretation. How does he proceed? How does he manage to introduce absurd and counterinductive assertions, such as the assertion that the earth moves, and yet get them a just and attentive hearing? One anticipates that arguments will not suffice – an interesting and highly important limitation of rationalism – and Galileo's utterances are indeed arguments in appearance only. For Galileo uses *propaganda*. He uses *psychological tricks* in addition to whatever intellectual reasons he has to offer. These tricks are very successful: they lead him to victory. But they obscure the new attitude towards experience that is in the making, and postpone for centuries the possibility of a reasonable philosophy. They obscure the fact that the experience on which Galileo wants to base the Copernican view is nothing but the result of his own fertile imagination, that it has been *invented*. They obscure this fact by insinuating that the new results which emerge are known and conceded by all, and need only be called to our attention to appear as the most obvious expression of the truth.

Galileo 'reminds' us that there are situations in which the non-operative character of shared motion is just as evident and as firmly believed as the idea of the operative character of all motion is in other circumstances. (This latter idea is, therefore, not the only natural interpretation of motion.) The situations are: events in a boat, in a smoothly moving carriage, and in other systems that contain an observer and permit him to carry out some simple operations.

'*Sagredo:* There has just occurred to me a certain fantasy which passed through my imagination one day while I was sailing to Aleppo, where I was going as consul for our country. . . . If the point of a pen had been on the ship during my whole voyage from Venice to Alexandretta and had had the property of leaving visible marks of its whole trip, what trace – what mark – what line would it have left?

Simplicio: It would have left a line extending from Venice to there; not perfectly straight – or rather, not lying in the perfect arc of a circle – but more or less fluctuating according as the vessel would now and again have rocked. But this bending in some places a yard or two to the right or left, up or down, in length of many hundreds of miles, would have made little alteration in the whole extent of the line. These would scarcely be sensible, and, without an error of any moment, it could be called part of a perfect arc.

Sagredo: So that if the fluctuation of the waves were taken away and the motion of the vessel were calm and tranquil, the true and precise motion of that pen point would have been an arc of a perfect circle. Now if I had had that same pen continually in my hand, and had moved it only a little sometimes this way or that, what alterations should I have brought into the main extent of this line?

Simplicio: Less than that which would be given to a straight line a thousand yards long which deviated from absolute straightness here and there by a flea's eye.

Sagredo: Then if an artist had begun drawing with that pen on a sheet of paper when he left the port and had continued doing so all the way to Alexandretta, he would have been able to derive from the pen's motion a whole narrative of many figures, completely traced and sketched in thousands of directions, with landscapes, buildings, animals, and other things. Yet the actual real essential movement marked by the pen point would have been only a line; long, indeed, but very simple. But as to the artist's own actions, these would have been conducted exactly the same as if the ship had been standing still. The reason that of the pen's long motion no trace would remain except the marks drawn upon the paper is that the gross motion from Venice to Alexandretta was common to the paper, the pen, and everything else in the ship. But the small motions back and forth, to right and left, communicated by the artist's fingers to the pen but not to the paper, and belonging to the former alone, could thereby leave a trace on the paper which remained stationary to those motions.'[1]

Or

'*Salviati:* ... imagine yourself in a boat with your eyes fixed on a point of the sail yard. Do you think that because the boat is moving along briskly, you will have to move your eyes in order to keep your vision always on that point of the sail yard and follow its motion?

Simplicio: I am sure that I should not need to make any change at all; not just as to my vision, but if I had aimed a musket I should never have to move it a hairsbreadth to keep it aimed, no matter how the boat moved.

Salviati: And this comes about because the motion which the ship confers upon the sail yard, it confers also upon you and upon your eyes, so that you need not move them a bit in order to gaze at the top of the sail yard, which consequently appears motionless to you. (And the rays of vision go from the eye to the sail yard just as if a cord were tied between the two ends of the boat. Now a hundred cords are tied at different fixed points, each of which keeps its place whether the ship moves or remains still.)'[2]

It is clear that these situations lead to a non-operative concept of motion even within common sense.

On the other hand, common sense, and I mean 17th-century Italian-artisan common sense, also contains the idea of the *operative* character of all motion. This latter idea arises when a limited object that does not contain too many parts moves in vast and stable surroundings; for

1. *Dialogue*, op. cit., pp. 171ff.

2. ibid., pp. 249ff. That phenomena of *seen* motion depend on *relative* motion has been asserted by Euclid in his *Optics*, Theon red. par. 49ff. An old scholion of par. 50 uses the example of a boat leaving the harbour: Heiberg, vii, 283. The example is repeated by Copernicus in Book I, Chapter viii, of *De Revol.* It was a commonplace in mediaeval optics. Cf. Witelo, *Perspectiva*, iv, par. 138 (Basel, 1572, p. 180). We know now that it is valid for constant velocities only.

example, when a camel trots through the desert, or when a stone descends from a tower.

Now Galileo urges us to 'remember' the conditions in which we assert the non-operative character of shared motion in this case also, and to subsume the second case under the first.

Thus, the first of the two paradigms of non-operative motion mentioned above is followed by the assertion that – 'It is likewise true that the earth being moved, the motion of the stone in descending is actually a long stretch of many hundred yards, or even many thousand; and had it been able to mark its course in motionless air or upon some other surface, it would have left a very long slanting line. But that part of all this motion which is common to the rock, the tower, and ourselves remains insensible and as if it did not exist. There remains observable only that part in which neither the tower nor we are participants; in a word, that with which the stone, in falling, measures the tower.'[3]

And the second paradigm precedes the exhortation to 'transfer this argument to the whirling of the earth and to the rock placed on top of the tower, whose motion you cannot discern because, in common with the rock, you possess from the earth that motion which is required for following the tower; you do not need to move your eyes. Next, if you add to the rock a downward motion which is peculiar to it and not shared by you, and which is mixed with this circular motion, the circular portion of the motion which is common to the stone and the eye continues to be imperceptible. The straight motion alone is sensible, for to follow that you must move your eyes downwards.'[4]

This is strong persuasion indeed.

Yielding to this persuasion, we now *quite automatically* start confounding the conditions of the two cases and become relativists. This is the essence of Galileo's trickery! As a result, the clash between Copernicus and 'the conditions affecting ourselves and those in the air above us'[5] dissolves into thin air, and we finally realize 'that all terrestrial events from which it is ordinarily held that the earth stands still and the sun and the fixed stars are moving would necessarily appear just the same to us if the earth moved and the other stood still'.[6]

3. ibid., pp. 172ff. 4. ibid., p. 250. 5. Ptolemy, *Syntaxis*, i, 1, p. 7.
6. *Dialogue*, p. 416: cf. the *Dialogues Concerning Two New Sciences*, transl. Henry

Let us now look at the situation from a more abstract point of view. We start with two conceptual sub-systems of 'ordinary' thought (see the following table). One of them regards motion as an absolute process which always has effects, effects on our senses included. The description of this conceptual system given here may be somewhat idealized; but the arguments of Copernicus' opponents which are quoted by Galileo himself and, according to him, are 'very plausible',[7] show that there was a widespread tendency to think in its terms, and that this tendency was a serious obstacle to the discussion of alternative ideas. Occasionally, one finds even more primitive ways of thinking, where concepts such as 'up' and 'down' are used absolutely. Examples are: the assertion 'that the earth is too heavy to climb up over the sun and then fall headlong back down again',[8] or the assertion that 'after a short time the mountains, sinking downward with the rotation of the terrestrial globe, would get into such a position that whereas a little earlier one would have had to climb steeply to their peaks, a few hours later one would have to stoop

---

Crew and Alfonso de Salvio, New York, 1958, p. 164: 'The same experiment which at first glance seemed to show one thing, when more carefully examined, assures us of the contrary.' Professor McMullin, in a critique of this way of seeing things, wants more 'logical and biographical justification' for my assertion that Galileo not only argued, but also cheated ['A Taxonomy of the Relation between History and Philosophy of Science', *Minnesota Studies*, Vol. 5, Minneapolis, 1971, p. 39], and he objects to the way in which I let Galileo introduce dynamical relativism. According to him 'what Galileo argues is that since his opponent *already* interprets observations made in such a context [movements on boats] in a 'relativistic' way, how can he consistently do otherwise in the case of observations made on the earth's surface?' (op. cit., p. 40). This is indeed how Galileo argues. But he argues so against an opponent who, according to him, 'feels a great repugnance towards recognizing this non-operative quality of motion among the things which share it in common' (*Dialogue*, op. cit., p. 171), who is convinced that a boat, apart from having relative motions, *has absolute positions and motions as well* (cf. Aristotle, *Physics*, 208b8ff), and who at any rate has developed the art of using different notions on different occasions without running into a contradiction. Now if *this* is the position to be attacked, then showing that an opponent has a relative idea of motion, or frequently uses the relative idea in his everyday affairs, is not at all 'proof of inconsistency in his own "paradigm" ' (McMullin, op. cit., p. 40). It just reveals one part of that paradigm without touching the other. The argument turns into the desired proof only if the absolute notion is either suppressed or spirited away, or else identified with the relativistic notion – and this is what Galileo actually does, though surreptitiously, as I have tried to show.

7. *Dialogue*, op. cit., p. 131.

8. ibid., p. 327.

and descend in order to get there'.[9] Galileo, in his marginal notes, calls these 'utterly childish reasons [which] sufficed to keep imbeciles believing in the fixity of the earth',[10] and he thinks it unnecessary 'to bother about such men as those, *whose name is legion*, or to take notice of their fooleries'.[11] Yet it is clear that the absolute idea of motion was 'well-entrenched', and that the attempt to replace it was bound to encounter strong resistance.[12]

The second conceptual system is built around the relativity of motion, and is also well-entrenched in its own domain of application. Galileo aims at replacing the first system by the second in *all* cases, terrestrial as well as celestial. Naive realism with respect to motion is to be *completely eliminated*.

9. ibid., p. 330. 10. ibid., p. 327.
11. ibid, p. 327, italics added.
12. The idea that there is an absolute direction in the universe has a very interesting history. It rests on the structure of the gravitational field on the surface of the earth, or of that part of the earth which the observer knows, and generalizes the experiences made there. The generalization is only rarely regarded as a separate hypothesis, it rather enters the 'grammar' of common sense and gives the terms 'up' and 'down' an absolute meaning. (This is a 'natural interpretation', in precisely the sense that was explained in the text above.) Lactantius, a church father of the fourth century, appeals to this meaning when he asks (*Divinae Institutiones*, III, De Falsa Sapientia): 'Is one really going to be so confused as to assume the existence of humans whose feet are above their heads? Where trees and fruit grow not upwards, but downwards?' The same use of language is presupposed by that 'mass of untutored men' who raise the question why the antipodeans are not falling off the earth (Pliny, *Natural History*, II, pp. 161–6; cf. also Ptolemy, *Syntaxis*, I, 7). The attempts of Thales, Anaximenes and Xenophanes to find support for the earth which prevents it from falling 'down' (Aristotle, *De Coelo*, 294a12ff) shows that almost all early philosophers, with the sole exception of Anaximander, shared in this way of thinking. (For the Atomists, who assume that the atoms originally fall 'down', cf. Jammer, *Concepts of Space*, Cambridge, Mass., 1953, p. 11.) Even Galileo, who thoroughly ridicules the idea of the falling antipodes (*Dialogue*, op. cit., p. 331), occasionally speaks of the 'upper half of the moon', meaning that part of the moon 'which is invisible to us'. And let us not forget that some linguistic philosophers of today 'who are too stupid to recognize their own limitations' (Galileo, op. cit., p. 327) want to revive the absolute meaning of 'up–down' at least *locally*. Thus the power over the minds of his contemporaries of a primitive conceptual frame, assuming an anisotropic world, which Galileo had also to fight, must not be underestimated. For an examination of some aspects of British common sense at the time of Galileo, including astronomical common sense, see E. M. W. Tillyard, *The Elizabethan World Picture*, London, 1963. The agreement between popular opinion and the centrally symmetric universe is frequently asserted by Aristotle, e.g. in *De Coelo*, p. 308a23f.

PARADIGM I: Motion of compact objects in stable surroundings of great spatial extension – deer observed by the hunter.		PARADIGM II: Motion of objects in boats, coaches and other moving systems.	
Natural interpretation: All motion is operative.		*Natural interpretation:* Only relative motion is operative.	
Falling stone *proves* ↓	Motion of earth *predicts* ↓	Falling stone *proves* ↓	Motion of earth *predicts* ↓
Earth at rest	Oblique motion of stone	No *relative* motion between starting point and earth	No relative motion between starting point and stone

Now, we have seen that this naive realism is on occasions an essential part of our observational vocabulary. On these occasions (Paradigm I), the observation language contains the idea of the efficacy of *all* motion. Or, to express it in the material mode of speech, our experience in these situations is an experience of objects which move absolutely. Taking this into consideration, it is apparent that Galileo's proposal amounts to a partial revision of our observation language or of our experience. An experience which partly *contradicts* the idea of the motion of the earth is turned into an experience that *confirms* it, at least as far as 'terrestrial things' are concerned.[13] This is what *actually happens*. But Galileo wants to persuade us that no change has taken place, that the second conceptual system is already universally *known*, even though it is not universally *used*. Salviati, his representative in the Dialogue, his opponent Simplicio and Sagredo the intelligent layman, all connect Galileo's method of argumentation with Plato's theory of *anamnesis* – a clever tactical move, typically Galilean one is inclined to say. Yet we must not allow ourselves to be deceived about the revolutionary development that is actually taking place.

13. *Dialogue*, op. cit., pp. 132 and 416.

The resistance against the assumption that shared motion is non-operative was equated with the resistance which forgotten ideas exhibit towards the attempt to make them known. Let us accept this *interpretation* of the resistance! But let us not forget its *existence*. We must then admit that it restricts the use of the relativistic ideas, confining them to *part* of our everyday experience. *Outside* this part, i.e. in interplanetary space, they are 'forgotten' and therefore not active. But outside this part there is not complete chaos. Other concepts are used, among them whose very same absolutistic concepts which derive from the first paradigm. We not only use them, but we must admit that they are entirely adequate. No difficulties arise as long as one remains within the limits of the first paradigm. 'Experience', i.e. the totality of all facts from all domains, cannot force us to carry out the change which Galileo wants to introduce. The motive for a change must come from a different source.

It comes, first, from the desire to see 'the whole [correspond] to its parts with wonderful simplicity',[14] as Copernicus had already expressed himself. It comes from the 'typically metaphysical urge' for unity of understanding and conceptual presentation. And the motive for a change is connected, secondly, with the intention to make room for the motion of the earth, which Galileo accepts and is not prepared to give up. The idea of the motion of the earth is closer to the first paradigm than to the second, or at least it was at the time of Galileo. This gave strength to the Aristotelian arguments, and made them plausible. To eliminate this plausibility, it was necessary to subsume the first paradigm under the second, and to extend the relative notions to all phenomena. The idea of *anamnesis* functions here as a psychological crutch, as a lever which smooths the process of subsumption by concealing its existence. As a result we are now *ready* to apply the relative notions not only to boats,

14. ibid., p. 341. Galileo quotes here from Copernicus' address to Pope Paul III in *De Revolutionibus*; cf. also Chapter 10 and the *Narratio Prima* (quoted from E. Rosen, *Three Copernican Treatises*, New York, 1959, p. 165): 'For all these phenomena appear to be linked most nobly together, as by a golden chain; and each of the planets, by its position, and order, and every inequality of its motion, bears witness that the earth moves and that we who dwell upon the globe of the earth, instead of accepting its changes of position, believe that the planets wander in all sorts of motions of their own.' Note that empirical reasons are absent from the argument and have to be, for Copernicus himself admits (*Commentariolus*, op. cit., p. 57) that the Ptolemaic theory is 'consistent with the numerical data'.

coaches, birds, but to the 'solid and well-established earth' as a whole. And we have the impression that this readiness was in us all the time, although it took some effort to make it conscious. This impression is most certainly erroneous: it is the result of Galileo's propagandistic machinations. We would do better to describe the situation in a different way, as a change of our conceptual system. Or, because we are dealing with concepts which belong to natural interpretations, and which are therefore connected with sensations in a very direct way, we should describe it as a *change of experience* that allows us to accommodate the Copernican doctrine. The change corresponds perfectly to the pattern described in Chapter 11 below: an inadequate view, the Copernican theory, is supported by another inadequate view, the idea of the non-operative character of shared motion, and both theories gain strength and give support to each other in the process. It is this change which underlies the transition from the Aristotelian point of view to the epistemology of modern science.

For experience now ceases to be the unchangeable fundament which it is both in common sense and in the Aristotelian philosophy. The attempt to support Copernicus makes experience 'fluid' in the very same manner in which it makes the heavens fluid, 'so that each star roves around in it by itself'.[15] An empiricist who starts from experience, and builds on it without ever looking back, now loses the very ground on which he stands. Neither the earth, 'the solid, well-established earth', nor the facts on which he usually relies can be trusted any longer. It is clear that a philosophy that uses such a fluid and changing experience needs new methodological principles which do not insist on an asymmetric judgement of theories by experience. *Classical physics* intuitively adopts such principles; at least the great and independent thinkers, such as Newton, Faraday, Boltzmann proceed in this way. But its *official doctrine* still clings to the idea of a stable and unchanging basis. The clash between this doctrine and the actual procedure is concealed by a tendentious presentation of the *results* of research that hides their revolutionary origin and suggests that they arose from a stable and unchanging source. These methods of concealment start with Galileo's attempt to introduce new ideas under the cover of anamnesis, and they

15. *Dialogue*, op. cit., p. 120.

culminate in Newton.[16] They must be exposed if we want to arrive at a better account of the progressive elements in science.

My discussion of the anti-Copernican argument is not yet complete. So far, I have tried to discover what assumption will make a stone *that moves alongside a moving tower* appear to fall 'straight down', instead of being seen to move in an arc. The assumption, which I shall call the *relativity principle*, that our senses notice only relative motion and are completely insensitive to a motion which objects have in common, was seen to do the trick. What remains to be explained is *why the stone stays with the tower* and is not left behind. In order to save the Copernican view, one must explain not only why a motion that preserves the relation among visible objects remains unnoticed, but also, why a common motion of various objects does not affect their relation. That is, one must explain why such a motion is not a *causal agent*. Turning the question around in the manner explained in text to footnote 10, page 78 of the last chapter, it is now apparent that the anti-Copernican argument described there rests on *two* natural interpretations: viz., the *epistemological assumption* that absolute motion is always *noticed*, and the *dynamical principle* that objects (such as the falling stone) which are not interfered with assume their natural motion. The present problem is to supplement the relativity principle with a new law of inertia in such a fashion that the motion of the earth can still be asserted. One sees at once that the following law, the *principle of circular inertia* as I shall call it, provides the required solution: an object that moves with a given angular velocity on a frictionless sphere around the centre of the earth will continue moving with the same angular velocity forever. Combining the appearance of the falling stone with the relativity principle, the principle of circular inertia and with some simple assumptions concerning the composition of velocities,[17] we obtain an argument which no longer endangers Copernicus' view, but can be used to give it partial support.

The relativity principle was defended in two ways. The first was by showing how it helps Copernicus: this defence is truly *ad hoc*. The second was by pointing to its function in common sense, and by surrep-

16. 'Classical Empiricism', op. cit.

17. These assumptions were not at all a matter of course, but conflicted with some very basic ideas of Aristotelian physics.

titiously generalizing that function (see above). No independent argument was given for its validity. Galileo's support for the principle of circular inertia is of exactly the same kind. He introduces the principle, again not by reference to experiment or to independent observation, but by reference to what everyone is already supposed to know.

'*Simplicio:* So you have not made a hundred tests, or even one? And yet you so freely declare it to be certain? . . .

Salviati: Without experiment, I am sure that the effect will happen as I tell you, because it must happen that way; and I might add that you yourself also know that it cannot happen otherwise, no matter how you may pretend not to know it. . . . But I am so handy at picking people's brains that I shall make you confess this in spite of yourself.'[18]

Step by step, Simplicio is forced to admit that a body that moves, without friction, on a sphere concentric with the centre of the earth will carry out a 'boundless', a 'perpetual' motion. We know, of course, especially after the analysis we have just completed of the non-operative character of shared motion, that what Simplicio accepts is based neither on experiment nor on corroborated theory. It is a daring new suggestion involving a tremendous leap of the imagination. A little more analysis then shows that this suggestion is connected with experiments, such as the 'experiments' of the *Discorsi*,[19] by *ad hoc* hypotheses. (The amount

18. *Dialogue*, op. cit., p. 147.

19. Incidentally, many of the 'experiences' or 'experiments' used in the arguments about the motion of the earth are entirely fictitious. Thus Galileo, in his *Trattato della Sfera* (*Edizione Nazionale*, Vol. II, pp. 211ff), which 'follows the opinion of Aristotle and of Ptolemy' (p. 223), uses this argument against a rotation of the earth: '. . . objects which one lets fall from high places to the ground such as a stone from the top of a tower would not fall towards the foot of that tower; for during the time which the stone coming rectilinearly towards the ground, spends in the air, the earth, escaping it, and moving towards the east would receive it in a part far removed from the foot of the tower *in exactly the same manner in which a stone that is dropped from the mast of a rapidly moving ship will not fall towards its foot, but more towards the stern*' (p. 224). The italicized reference to the behaviour of stones on ships is again used in the *Dialogue* (p. 126), when the Ptolemaic arguments are discussed, but it is no longer accepted as correct. 'It seems to be an appropriate time,' says Salviati (ibid., p. 180), 'to take notice of a certain generosity on the part of the Copernicans towards their adversaries when, with perhaps too much liberality, they concede as true and correct a number of experiments which their opponents have never made. Such for example is that of the body

of friction to be eliminated follows not from independent investigations – such investigations commence only much later, in the 18th century – but from the result to be achieved, viz. the circular law of inertia.) Viewing natural phenomena in this way leads to a re-evaluation of all experience, as we have seen. We can now add that it leads to the invention of a *new kind of experience* that is not only more sophisticated *but also far more speculative* than is the experience of Aristotle or of common sense. Speaking paradoxically, but not incorrectly, one may say that *Galileo invents an experience that has metaphysical ingredients.* It is by means of such an experience that the transition from a geostatic cosmology to the point of view of Copernicus and Kepler is achieved.

---

falling from the mast of a ship while it is in motion. . . .' Earlier, p. 154, it is implied rather than observed, that the stone will fall to the foot of the mast, even if the ship should be in motion while a possible experiment is discussed on p. 186. Bruno (*La Cena de le Ceneri*, *Opere Italiane*, I, ed. Giovanni Gentile, Bari, 1907, p. 83) takes it for granted that the stone will arrive at the foot of the mast. It should be noted that the problem did not readily lend itself to an experimental solution. Experiments were made, but their results were far from conclusive. Cf. A. Armitage, 'The Deviation of Falling Bodies', *Annals of Science*, 5, 1941–7, pp. 342ff, and A. Koyré, *Metaphysics and Measurement*, Cambridge, 1968, pp. 89ff. The tower argument can be found in Aristotle, *De Coelo*, 296b22, and Ptolemy, *Syntaxis*, i, 8. Copernicus discusses it in the same chapter of *De Revol.*, but tries to defuse it in the next chapter (cf. footnote 12 to Chapter 8 of the present essay). Its role in the middle ages is described in M. Clagett, *The Science of Mechanics in the Middle Ages*, Madison, 1959, Chapter 10.

8

Initial difficulties caused by the change are defused by ad hoc hypotheses, *which thus turn out occasionally to have a positive function; they give new theories a breathing space, and they indicate the direction of future research.*

This is the place to mention certain ideas developed by Lakatos which throw new light on the problem of the growth of knowledge and which, to some extent, undermine his own quest for law and order in science.

It is customary to assume that good scientists refuse to employ *ad hoc* hypotheses and are right to do so. New ideas, so it is thought, go far beyond the available evidence and *must* go beyond it in order to be of value. *Ad hoc* hypotheses are bound to creep in eventually, but they should be resisted and kept at bay. This is the customary attitude as it is expressed, for example, in the writings of K. R. Popper.

As opposed to this, Lakatos has pointed out that 'adhocness' is neither despicable, nor absent from the body of science.[1] New ideas, he emphasizes, are usually almost entirely *ad hoc*, they cannot be otherwise. And they are reformed only in a piecemeal fashion, by gradually stretching them, so that they apply to situations lying beyond their starting point. Schematically:

Popper: New theories have, and must have, excess content which is, but should not be, gradually infected by *ad hoc* adaptations.

1. Cf. Lakatos in *Criticism and the Growth of Knowledge*, Cambridge, 1970. The use of *ad hoc* hypotheses in the sciences is identical with what anthropologists have called 'secondary elaborations'. (See R. Horton, 'African Traditional Thought and Western Science' in *Witchcraft and Sorcery*, ed. N. Marwick, London, 1970. p. 35.) Secondary elaborations are supposed to be one *differentia specifica* separating science from witchcraft. Our considerations in the text (and in Chapter 12 below) refute this assumption and show that differences – if there are differences – must be looked for elsewhere.

Lakatos: New theories are, and cannot be anything but, *ad hoc*. Excess content is, and should be, created in a piecemeal fashion, by gradually extending them to new facts and domains.

The historical material I have been discussing (and the material in Chapters 9–11 below) lends unambiguous support to the position of Lakatos. The early history of Galileo's mechanics tells exactly the same story.

In *De Motu*[2] motions of spheres in the centre of the universe and outside it, homogeneous and non-homogeneous, supported at the centre of gravity and supported outside it, are discussed and described as being either natural, or forced, or neither. But we hear very little about the actual motion of such spheres at this place, and what we do hear is by implication only. For example, there arises the question whether a homogeneous sphere made to move in the centre of the universe would move forever.[3] We read that 'it seems that it should move perpetually', but an unambiguous answer is not given. A marble sphere supported on an axis through the centre and set in motion is said to 'rotate for a long time',[4] in *De Motu*; while in the *Dialogue on Motion*, a perpetual motion is said to be 'quite out of keeping with the nature of the earth itself, to which rest seems to be more congenial than motion'.[5] Another, more specific argument against perpetual rotations is found in Benedetti's *Diverse Speculations*.[6] Rotations, says Benedetti, are 'certainly not perpetual', for the parts of the sphere, wanting to move in a straight line, are constrained against their nature, 'and so they come to rest naturally'. Again, in *De Motu* we find a criticism of the assertion that adding a star to the celestial sphere might slow the sphere down by changing the relation between the force of the moving intelligences and its resistance.[7] This assertion, Galileo says, applies to an excentric sphere. Adding a weight to an excentric sphere means that the weight will occasionally move away from the centre and be raised to a higher level. But 'who

2. Galileo Galilei, *De Motu*. Quoted from *Galileo Galilei on Motion and on Mechanics*, ed. Drake and Drabkin, Madison, 1960, p. 73.

3. ibid., p. 73. 4. ibid., p. 78.

5. Quoted from *Mechanics in Sixteenth Century Italy*, ed. Drake and Drabkin, Madison, 1969, p. 338. In footnote 10 on the same page, Drake comments that 'Galileo was not a Copernican when he wrote this'.

6. ibid., p. 228. 7. ibid., pp. 73ff.

would ever say that a concentric sphere was impeded by the weight, since the weight in its circular path would neither approach, nor recede from the centre'.[8] Note that the original rotation is, in this case, said to be caused by the 'intelligences'; it is not assumed to be taking place by itself. This is in perfect agreement with Aristotle's *general* theory of motion, where a mover is postulated for *every* motion and not just for violent ones.[9] Galileo seems to accept this part of the theory, both when letting rotating spheres slow down and when accepting the 'force of the intelligences'. He also accepts the impetus theory which attributes *any* motion to an internal moving force similar to the force of sound that resides in a bell long after it has been struck,[10] and is supposed to 'gradually diminish'.[11]

Looking at these few examples, we see that Galileo ascribes a special position to motions which are neither natural nor forced. Such motions may last for a considerable time, even though they are not supported by the surrounding medium. But *they do not last forever*, and *they need an internal driving force* in order to persist even for a finite time.

Now, if one wants to overcome the dynamical arguments against the motion of the earth (and we are here thinking about its *rotation* rather than about its motion around the sun), then the two underlined principles must be both revised. It must be assumed that the 'neutral' motions, which Galileo discusses in his early dynamical writings, may last forever, or at least for periods comparable to the age of historical records. And they must be regarded as 'natural', in the entirely new and revolutionary sense that neither an outer *nor an inner motor* is needed to keep them going. The first assumption is necessary to account for the phenomenon of the daily rising and setting of the stars. The second assumption is necessary if we want to regard motion as a *relative* phenomenon, depending on the choice of a suitable co-ordinate system. Copernicus, in his brief remarks on the problem, makes the first assumption, and perhaps also the second.[12] Galileo takes a long time arriving at

8. ibid., p. 74.
9. *Physics*, VII, 1, 241b34–6.
10. *De Motu*, op. cit., p. 79.
11. *De Motu*, op. cit., viii (in Drabkin's sub-division).
12. *De Revolutionibus*, I, Chapter 8: 'Circular motion however is always [of constant velocity] *because it has a non-ceasing cause*' (my italics). Copernicus accepts the Aristotelian theory of motion and of the elements and he tries to account for the rotation of

a comparable theory. He formulates permanence along a horizontal line as a hypothesis in his *Discorsi*,[13] and he seems to make both assumptions in the *Dialogue*.[14] My guess is that *a clear idea of permanent motion with(out) impetus developed in Galileo only together with his gradual acceptance of the Copernican view*. Galileo changed his view about the 'neutral' motions – he made them permanent and 'natural' – in order to make them compatible with the rotation of the earth and in order to evade the difficulties of the tower argument.[15] His new ideas concerning

---

the earth in its terms. The reference to a 'cause' is ambiguous. It might imply a version of the impetus theory, but it might simply mean that the earth rotates with constant angular velocity because it resides in its natural place: 'Hence, a simple body has a simple motion which is mainly shown in the case of circular motion, as long as the simple body resides in its natural place and preserves its unity. In that place motion cannot be but circular, which remains entirely in itself, as if the body remained at rest.' Considering that Copernicus regards the classification of motions into straight and circular as a mathematical artifice 'comparable to the way in which we distinguish line, point, and surface while the one cannot exist without the other, and while none can exist without a body' the second interpretation seems to be preferable (though, regarding the world as an 'animal' he still assumes absolute space – see below). For these problems cf. the remarks of Birkenmajer in footnote 82ff of G. Klaus (ed.), *Copernicus über Kreisbewegung*, Berlin, 1959. Cf. also the third dialogue of Bruno's *La Cena de le Ceneri*, op. cit., pp. 76–85, especially 82ff. The principle used by Bruno (and, perhaps, also by Copernicus) that the earth is an *organism* whose parts are bound to move with the whole, may have been taken from the *Discourse of Hermes to Tat* (English transl. in Scott, *Hermetica*, Vol. I). Copernicus mentions Hermes once, in *De Revol.*, i, 10, discussing the position of the sun: 'But in the centre there rests the sun . . . whom Trimegistus [sic] calls the visible god . . .', cf. footnote 5 to Chapter 4. He likens the *world* to an organism in which circular motion coexists with rectilinear motion just as the organism coexists with its sickness. (This whole problem of the relation between straight motion and circular motion is discussed at length in the *First Day* of Galileo's *Dialogue*.) The *earth*, however, 'conceives from the sun and becomes pregnant with yearly birth' (Chapter 10). For a survey of reactions towards the physical difficulties of the motion of the earth cf. Chapter 1 of Vol. III of A. Koyré's *Etudes Galiléennes*, Paris, 1939.

13. *Two New Sciences*, New York, 1954, pp. 215 and 250.

14. op. cit., pp. 147ff.

15. According to Anneliese Maier (*Die Vorläufer Galileis im 14. Jahrhundert*, Rome, 1949, pp. 151ff), Galileo replaced impetus by inertia in order to explain the 'fact' that 'neutral' motions go on forever. To start with, there was no such 'fact'. Secondly, Galileo initially did not believe, and rightly so, that there was such a fact. This we have just seen. There was, therefore, no need for him 'to explain certain newly detected *phenomena*' (p. 151). The need was purely theoretical: to accommodate, to 'save', not a phenomenon *but a new world view*. For the insufficiency of contemporary experiments

such motions are, therefore, at least partly *ad hoc*. Impetus in the old sense disappeared, partly for methodological reasons (interest in the *how*, not in the *why* – this development itself deserves careful study), partly because of the vaguely perceived inconsistency with the idea of the relativity of all motion. The wish to save Copernicus plays a role in either case.

Now, if we are right in *assuming* that Galileo framed an *ad hoc* hypothesis at this point, then we can also *praise* him for his methodological acumen. It is obvious that the moving earth demands a new dynamics. *One* test of the old dynamics consists in the attempt to establish the motion of the earth. Trying to establish the motion of the earth is the same as trying to find a refuting instance for the old dynamics. The motion of the earth, however, is inconsistent with the tower experiment *if this experiment is interpreted in accordance with the old dynamics.* Interpreting the tower experiment in accordance with the old dynamics, therefore, means trying to save the old dynamics in an *ad hoc* fashion. If one does not want to do this one must find a different interpretation for the phenomena of free fall. What interpretation should be chosen? One wants an interpretation that turns the motion of the earth into a refuting instance of the old dynamics, without lending *ad hoc* support to the motion of the earth itself. The first step towards such an interpretation is to establish contact, however vague, with the 'phenomena', i.e. with the falling stone, and to establish it in such a manner that the motion of the earth is not *obviously* contradicted. The most primitive element of this step is to frame an *ad hoc* hypothesis with respect to the rotation of the earth. The next step would be to elaborate the hypothesis, so that additional predictions become possible. Copernicus and Galileo take the

---

cf. footnote 19 of the preceding chapter. Stillman Drake has asserted, in a most interesting and provocative essay, that 'Galileo as a physicist, treated inertial motions as rectilinear. Nevertheless, Galileo as a propagandist, when writing the *Dialogue*, stated that rectilinear motion cannot be perpetual, though circular motion may be. . . . Accordingly, when I read the metaphysical praise of circles in the *Dialogue*, I do not conclude with most historians that its author was unable to break the spell of ancient traditions; rather, I strongly suspect an ultimate purpose in those passages' (*Galileo Studies*, Ann Arbor, 1970, p. 253). For this assertion he brings a great number of most convincing arguments. All this, of course, fits in quite marvellously with the ideology of the present essay.

first and most primitive step. Their procedure looks contemptible only if one forgets that the aim is *to test older views* rather than *to prove new ones*, and if one also forgets that developing a good theory is a complex process that has to start modestly and that takes time. It takes time *because the domain of possible phenomena must first be circumscribed by the further development of the Copernican hypothesis.* It is much better to remain *ad hoc* for a while, and in the meantime to develop heliocentrism in all its astronomical ramifications, than to sink back into the earlier ideas which, at any rate, can be defended only with the help of *other ad hoc* hypotheses.

Therefore, Galileo *did* use *ad hoc* hypotheses. It *was good* that he used them. Had he not done so, he would have been *ad hoc* anyway, but this time with respect to an older theory. Hence, since one cannot help being *ad hoc*, it is better to be *ad hoc* with respect to a new theory, for a new theory, like all new things, will give a feeling of freedom, excitement and progress. Galileo is to be applauded because he preferred protecting an interesting hypothesis to protecting a dull one.

9

In addition to natural interpretations, Galileo also changes sensations *that seem to endanger Copernicus. He admits that there are such sensations, he praises Copernicus for having disregarded them, he claims to have removed them with the help of his* telescope. *However, he offers no* theoretical *reasons why the telescope should be expected to give a true picture of the sky.*

I repeat and summarize. An argument is proposed that refutes Copernicus by observation. The argument is inverted in order to discover the natural interpretations which are responsible for the contradiction. The offensive interpretations are replaced by others, propaganda and appeal to distant, and highly theoretical, parts of common sense are used to defuse old habits and to enthrone new ones. The new natural interpretations, which are also formulated explicitly, as auxiliary hypotheses, are established partly by the support they give to Copernicus and partly by plausibility considerations and *ad hoc* hypotheses. An entirely new 'experience' arises in this way. Independent evidence is as yet entirely lacking, but this is no drawback as it is to be expected that independent support will take a long time appearing. For what is needed is a theory of solid objects, aerodynamics, and all these sciences are still hidden in the future. *But their task is now well-defined*, for Galileo's assumptions, his *ad hoc* hypotheses included, are sufficiently clear and simple to prescribe the direction of future research.

Let it be noted, incidentally, that Galileo's procedure drastically reduces the content of dynamics. Aristotelian dynamics was a general theory of change, comprising locomotion, qualitative change, generation and corruption, and it provided a theoretical basis for the theory of witchcraft also. Galileo's dynamics and its successors deal with *locomotion* only, and here again just with the locomotion of *matter*. Other kinds of motion are pushed aside with the promissory note (due to Democritos) that locomotion will eventually be capable of explaining all motion. Thus a comprehensive empirical theory of motion is replaced by a much narrower theory plus a metaphysics of motion,[1] just as an 'empirical' experience is replaced by an experience that contains speculative elements. *Counterinduction*, however, is now seen to play an important role both *vis-à-vis* theories and *vis-à-vis* facts. It clearly aids the advancement of science. This concludes the considerations begun in Chapter 6. I now turn to another part of Galileo's propaganda campaign, dealing not with natural interpretations but with the *sensory core* of our observational statements.

Replying to an interlocutor who expressed his astonishment at the small number of Copernicans, Salviati, who 'act(s) the part of Copernicus'[2] gives the following explanation: 'You wonder that there are so few followers of the Pythagorean opinion [that the earth moves] while I am

1. The so-called scientific revolution led to astounding discoveries and considerably extended our knowledge of physics, physiology, and astronomy. This was achieved by pushing aside and regarding as irrelevant, *and often as non-existent*, those facts which had supported the older philosophy. Thus all the evidence for witchcraft, demonic possession, the existence of the devil, etc., was disregarded *together with* the 'superstitions' it once confirmed. The result was that 'towards the close of the Middle Ages science was forced away from human psychology, so that even the great endeavour of Erasmus and his friend Vives, as the best representatives of humanism, did not suffice to bring about a reapproachment, and psychopathology had to trail centuries behind the developmental trend of general medicine and surgery. As a matter of fact, . . . the divorcement of medical science from psychopathology was so definite that the latter was always totally relegated to the domain of theology and ecclesiatic and civil law – two fields which naturally became further and further removed from medicine. . . .' G. Zilboorg, M.D., *The Medical Man and the Witch*, Baltimore, 1935, pp. 3ff and 70ff. Astronomy advanced, but the knowledge of man slipped back into an earlier and more primitive stage. Another example is astrology. 'In the early stages of the human mind,' writes A. Comte (*Cours de Philosophie Positive*, Vol. III, pp. 273–80, ed. Littré, Paris, 1836), 'these connecting links between astronomy and biology were studied from a very different point of view, *but at least* they were studied and not left out of sight, as is the common tendency in our own time, under the restricting influence of a nascent and incomplete positivism. Beneath the chimerical belief of the old philosophy in the physiological influence of the stars, there lay a strong, though confused recognition of the truth that the facts of life were in some way dependent on the solar system. Like all primitive inspirations of man's intelligence this feeling needed rectification by positive science, but not destruction; though unhappily in science, as in politics, it is often hard to reorganize without some brief period of overthrow.'

2. *Dialogue*, op. cit., pp. 131 and 256.

astonished that there have been any up to this day who have embraced and followed it. Nor can I ever sufficiently admire the outstanding acumen of those who have taken hold of this opinion and accepted it as true: they have, through sheer force of intellect done such violence to their own senses as to prefer what reason told them over that which sensible experience plainly showed them to be the contrary. For the arguments against the whirling [the rotation] of the earth we have already examined [the dynamical arguments discussed above] are very plausible, as we have seen; and the fact that the Ptolemaics and the Aristotelians and all their disciples, took them to be conclusive is indeed a strong argument of their effectiveness. But the experiences which overtly contradict the annual movement [the movement of the earth around the sun] are indeed so much greater in their apparent force that, I repeat, there is no limit to my astonishment when I reflect that Aristarchus and Copernicus were able to make reason so conquer sense that, in defiance of the latter, the former became mistress of their belief.'[3]

A little later Galileo notes that 'they [the Copernicans] were confident of what their reason told them!'[4] And he concludes his brief account of the origins of Copernicanism by saying that 'with reason as his guide he [Copernicus] resolutely continued to affirm what sensible experience seemed to contradict'. 'I cannot get over my amazement,' Galileo repeats, 'that he was constantly willing to persist in saying that Venus might go around the sun and might be more than six times as far from us at one time as at another, and still look always equal, when it should have appeared forty times larger.'[5]

The 'experiences which overtly contradict the annual movement', and which 'are much greater in their apparent force' than even the dynamical arguments above, consist in the fact that 'Mars, when it is close to us . . . would have to look sixty times as large as when it is most distant. Yet no such difference is to be seen. Rather, when it is in opposition to the sun and close to us it shows itself only four or five times as large as when, at conjunction, it becomes hidden behind the rays of the sun.'[6]

3. ibid., p. 328. At other times Galileo speaks much more belligerently and dogmatically, and apparently without any awareness of the difficulties mentioned here. Cf. his preparatory notes for the letter to Grand Duchess Christina, *Opera*, V, pp. 367ff.

4. ibid., p. 335. 5. ibid., p. 339. 6. ibid., p. 334.

'Another and greater difficulty is made for us by Venus which, if it circulates around the sun, as Copernicus says, would now be beyond it and now on this side of it, receding from and approaching towards us by as much as the diameter of the circle it describes. Then, when it is beneath the sun and very close to us, its disc ought to appear to us a little less than forty times as large as when it is beyond the sun and near conjunction. Yet the difference is almost imperceptible.'[7]

In an earlier essay, *The Assayer*, Galileo expressed himself still more bluntly. Replying to an adversary who had raised the issue of Copernicanism he remarks that '*neither Tycho, nor other astronomers nor even Copernicus could clearly refute* (*Ptolemy*) inasmuch as a most important argument taken from the movement of Mars and Venus stood always in their way'. (This 'argument' is mentioned again in the *Dialogue*, and has just been quoted.) He concludes that 'the two systems' (the Copernican and the Ptolemaic) are 'surely false'.[8]

We see again that Galileo's view of the origin of Copernicanism differs markedly from the more familiar historical accounts. He neither points to *new facts* which offer inductive *support* to the idea of the moving earth, nor does he mention any observations that would *refute* the geocentric point of view but be accounted for by Copernicanism. On the contrary, he emphasizes that not only Ptolemy, but Copernicus as well, is refuted by the facts,[9] and he praises Aristarchus and Copernicus for not having

7. For details concerning the study of the variation of planetary magnitudes cf. Appendix 1 to the present chapter.

8. *The Assayer*, quoted from *The Controversy on the Comets of 1918*, op. cit., p. 184.

9. This refers to the period before the end of the 16th century; cf. Derek J. de S. Price, 'Contra-Copernicus: A Critical Re-Estimation of the Mathematical Planetary Theory of Ptolemy, Copernicus and Kepler', *Critical Problems in the History of Science*, ed. M. Clagett, Madison, 1959, pp. 197–218. Price deals only with the *kinematic* and the *optical* difficulties of the new views. (A consideration of the dynamical difficulties would further strengthen his case.) He points out that 'under the best conditions a geostatic or heliostatic system using eccentric circles (or their equivalents) with central epicycles can account for all angular motions of the planets to an accuracy better than 6′ . . . excepting only the special theory needed to account for . . . Mercury and excepting also the planet Mars which shows deviations up to 30′ from such a theory. [This is] certainly better than the accuracy of 10′ which Copernicus himself stated as a satisfactory goal for his own theory' which was difficult to test, especially in view of the fact that refraction (almost 1° on the horizon) was not taken into account at the time of Copernicus, and that the observational basis of the predictions was less than satisfactory.

given up in the face of such tremendous difficulties. He praises them for having proceeded *counterinductively*.

This, however, is not yet the whole story.[10]

For while it might be admitted that Copernicus acted simply on faith,[11] it may also be said that Galileo found himself in an entirely different position. Galileo, after all, invented a new dynamics. And he invented the telescope. The new dynamics, one might want to point out, removes the inconsistency between the motion of the earth and the 'conditions affecting ourselves and those in the air above us'.[12] And the telescope removes the 'even more glaring' clash between the changes in the apparent brightness of Mars and Venus as predicted on the basis of the Copernican scheme and as seen with the naked eye. This, incidentally, is also Galileo's own view. He admits that 'were it not for the existence of a superior and better sense than natural and common sense to join forces with reason' he would have been 'much more recalcitrant towards the Copernican system'.[13] The 'superior and better sense' is, of course, the *telescope*, and one is inclined to remark that the apparently counterinductive procedure was as a matter of fact induction (or conjecture plus refutation plus new conjecture), *but one based on a better experience*, containing not only better natural interpretations but also a better sensory core than was available to Galileo's Aristotelian predecessors.[14] This matter must now be examined in some detail.

Carl Schumacher (*Untersuchungen über die ptolemäische Theorie der unteren Planeten*, Münster, 1917) has found that the predictions concerning Mercury and Venus made by Ptolemy differ at most by an amount of 30′ from those of Copernicus. The deviations found between modern predictions and those of Ptolemy (and Copernicus), which in the case of Mercury may be as large as 7°, are due mainly to wrong constants and initial conditions, including an incorrect value of the constant of precession. For the versatility of the Ptolemaic scheme cf. N. R. Hanson, *Isis*, No. 51, 1960, pp. 150–8.

10. Some historical statements made in this chapter and in the following chapters up to and including Chapter 11, and the inferences drawn from them are contested in a recent essay in *Studies in the History and Philosophy of Science*, May 1973, pp. 11–46, which was concocted by P. K. Machamer with the help of G. Buchdahl, L. Laudan and other experts. A discussion of the essay is found in Appendix 2 of the present Chapter.

11. He did not, as may be seen from footnote 12, Chapter 8, and footnote 7, Chapter 9 of the present essay.

12. Ptolemy, *Syntaxis*, i, 7.

13. *Dialogue*, op. cit., p. 328.

14. For this view cf. Ludovico Geymonat, *Galileo Galilei*, transl. Stillman Drake, New York, 1965 (first Italian edition 1957), p. 184.

The telescope is a 'superior and better sense' that gives new and more reliable evidence for judging astronomical matters. How is this hypothesis examined, and what arguments are presented in its favour?

In the *Sidereus Nuncius*,[15] the publication which contains his first telescopic observations, and which was also the first important contribution to his fame, Galileo writes that he 'succeeded (in building the telescope) through a deep study of the theory of refraction'. This suggests that he had *theoretical reasons* for preferring the results of telescopic observations to observations with the naked eye. But the particular reason he gives – his insight into the theory of refraction – is not *correct* and is not *sufficient* either.

The reason is not correct, for there exists serious doubts as to Galileo's knowledge of those parts of contemporary physical optics which were relevant for the understanding of telescopic phenomena. In a letter to Giuliano de Medici of 1 October 1610,[16] more than half a year after publication of the *Sidereus Nuncius*, he asks for a copy of Kepler's *Optics* of 1604,[17] pointing out that he had not yet been able to obtain it in Italy. Jean Tarde, who in 1614 asked Galileo about the construction of telescopes of pre-assigned magnification, reports in his diary that Galileo regarded the matter as a difficult one and that he had found Kepler's *Optics* of 1611[18] so obscure 'that perhaps its own author had

15. *The Sidereal Messenger of Galileo Galilei*, transl. E. St Carlos, London, 1880, reissued by Dawsons of Pall Mall, 1960, p. 10.

16. Galileo, *Opere*, Edit. Naz., x, p. 441.

17. *Ad Vitellionem Paralipomena quibus Astronomiae Pars Optica Traditur*, Frankfurt, 1604, to be quoted from *Johannes Kepler, Gesammelte Werke*, Vol. II, Munich, 1939, ed. Franz Hammer. This particular work will be referred to as the 'optics of 1604'. It was the only useful optics that existed at the time. The reason for Galileo's curiosity were most likely the many references to this work in Kepler's reply to the *Sidereus Nuncius*. For the history of this reply as well as a translation cf. *Kepler's Conversation with Galileo's Sidereal Messenger*, transl. E. Rosen, New York, 1965. The many references to earlier work contained in the *Conversation* were interpreted by some of Galileo's enemies as a sign that 'his mask had been torn from his face' (G. Fugger to Kepler, 28 May 1610, Galileo, *Opere*, Vol. X, p. 361) and that he (Kepler) 'had well plucked him', Maestlin to Kepler, 7 August (Galileo, *Opere*, Vol. X, p. 428). Galileo must have received Kepler's *Conversation* before 7 May (*Opere*, X, p. 349) and he acknowledges receipt of the printed *Conversation* in a letter to Kepler of 19 August (*Opere*, X, p. 421).

18. *Dioptrice*, Augsburg, 1611, *Werke*, Vol. IV, Munich, 1941. This work was written after Galileo's discoveries. Kepler's reference to them in the preface has been translated

not understood it'.[19] In a letter to Liceti, written two years before his death, Galileo remarks that as far as he was concerned the nature of light was still in darkness.[20] Even if we consider such utterances with the care that is needed in the case of a whimsical author, like Galileo, we must yet admit that his knowledge of optics was inferior by far to that of Kepler.[21] This is also the conclusion of Professor E. Hoppe, who sums up the situation as follows:

'Galileo's assertion that having heard of the Dutch telescope he reconstructed the apparatus by mathematical calculation must of course be understood with a grain of salt; for in his writings we do not find any calculations and the report, by letter, which he gives of his first effort says that no better lenses had been available; six days later we find him on the way to Venice with a better piece to hand it as a gift to the Doge Leonardi Donati. This does not look like calculation; it rather looks like trial and error. The calculation may well have been of a different kind, and here it succeeded, for on 25 August 1609 his salary was increased by a factor of three.'[22]

---

by E. St Carlos, op. cit., pp. 37, 79ff. The problem referred to by Tarde is treated in Kepler's *Dioptrice*.

19. Geymonat, op. cit., p. 37.

20. Letter to Liceti of 23 June 1640. *Opere*, VIII, p. 208.

21. Kepler, the most knowledgeable and most lovable of Galileo's contemporaries, gives a clear account of the reasons why, despite his superior knowledge of optical matters, he 'refrained from attempting to construct the device'. 'You, however,' he addresses Galileo, 'deserve my praise. Putting aside all misgivings you turned directly to visual experimentation' (*Conversation*, op. cit., p. 18). It remains to add that Galileo, due to his lack of knowledge in optics, had no 'misgivings' to overcome: 'Galileo . . . was totally ignorant of the science of optics, and it is not too bold to assume that this was a most happy accident both for him and for humanity at large', Ronchi, *Scientific Change*, ed. Crombie, London, 1963, p. 550.

22. *Die Geschichte der Optik*, Leipzig, 1926, p. 32. Hoppe's judgement concerning the invention of the telescope is shared by Wolf, Zinner and others. Huyghens points out that superhuman intelligence would have been needed to invent the telescope on the basis of the available physics and geometry. After all, says he, we still do not understand the workings of the telescope. ('Dioptrica', *Hugenii Opuscula Postuma*, Ludg. Bat., 1903, 163, paraphrased after A. G. Kästner, *Geschichte der Mathematik*, Vol. IV, Göttingen, 1800, p. 60.)

Various writers, whose lack of imagination and temperament is properly matched by their high moral standards, have been put off by the numerous signs of worldliness on the part of Galileo, and they have tried their best to explain his actions as the result

Trial and error – this means that 'in the case of the telescope it was *experience* and not mathematics that led Galileo to a serene faith in the reliability of his device'.[23] This second hypothesis on the origin of the telescope is *also* supported by Galileo's testimony, in which he writes that he had tested the telescope 'a hundred thousand times on a hundred

---

of high (and dry) motives. A much less important episode, viz. Galileo's silence about the achievements of Copernicus in his *Trattato della Sfera* (*Opere*, II, 211ff – the idea of the motion of the earth is mentioned, but not the name of Copernicus) at a time when, according to some, he had already accepted the Copernican creed has led to much soul searching and to some convenient *ad hoc* hypotheses even on the part of so worldly an author as L. Geymonat (op. cit., 23). And yet there is no reason why a man, and an extremely intelligent man at that, should conform to the standards of the academic squares of today and why he should not try in his own way to further his interests. It is a strange moral principle indeed that requires a thinker to be a blabbermouth who 'expresses' only what he believes to be 'the truth' and who never mentions what he does not believe. (Is this what the contemporary search for authenticity demands?) A Puritanical view like this surely is too naive a background for understanding a man of the late Renaissance and the early Baroque. Moreover Galileo the mountebank is a much more interesting character than the constipated 'searcher for the truth' we are usually invited to revere. *Finally, it was only through sleights of hand such as these that progress could be made at this particular time*, as we shall see. Cf. also footnote 19 of the present chapter.

Galileo's propaganda machinations are often guided by the insight that established institutions, social conditions, prejudices may hinder the acceptance of new ideas and that new ideas may therefore have to be introduced in an 'indirect' manner, by forging links between the circumstances of their origin and the forces that might endanger their survival. Doing this in the case of the Copernican doctrine Galileo more than once strays from the straight path of truth (whatever *that* is). In his letter to the Grand Duchess Christina (quoted after St. Drake, *Discoveries and Opinions of Galileo*, New York, 1957, p. 178) he says that 'Copernicus . . . was not only a Catholic, but a priest and a canon. He was in fact so esteemed by the church that when the Lateran Council under Leo X took up the correction of the Church calendar, Copernicus was called to Rome from the most remote parts of Germany to undertake its reform.' Actually, he never took orders, he was not called to Rome, and the Gregorian calendar decided against Copernicus. 'Then why did Galileo falsify this aspect of Copernicus's biography? As a loyal Catholic, Galileo was engaged in a valiant attempt to save his Church from committing the grave blunder [?] of condemning Copernicanism as a heresy. In the course of this hectic campaign Galileo made a number of historical misstatements about Copernicus, all intended to bind the revolutionary astronomer more closely to the Roman Catholic Church than the facts themselves warranted.' Rosen, biography of Copernicus in *Three Copernican Treatises*, New York, 1971, p. 320. This reminds us of Kant's observation that lies 'for the time being (may) have the function to raise mankind above its crude past', *Critique*, B 776, 15.

23. Geymonat, op. cit., p. 39.

thousand stars and other objects'.[24] Such tests produced great and surprising successes. The contemporary literature – letters, books, gossip columns – testifies to the extraordinary impression which the telescope made as a means of improving *terrestrial vision.*

Julius Caesar Lagalla, Professor of Philosophy in Rome, describes a meeting of 16 April 1611, at which Galileo demonstrated his device: 'We were on top of the Janiculum, near the city gate named after the Holy Ghost, where once is said to have stood the villa of the poet Martial, now the property of the Most Reverend Malvasia. By means of this instrument, we saw the palace of the most illustrious Duke Altemps on the Tuscan Hills so distinctly that we readily counted its each and every window, even the smallest; and the distance is sixteen Italian miles. From the same place we read the letters on the gallery, which Sixtus erected in the Lateran for the benedictions, so clearly, that we distinguished even the periods carved between the letters, at a distance of at least two miles.'[25]

24. Letter to Carioso, 24 May 1616, *Opere*, X, p. 357: letter to P. Dini, 12 May 1611, *Opere*, IX, p. 106: 'Nor can it be doubted that I, over a period of two years now, have tested my instrument (or rather dozens of my instruments) on hundreds and thousands of objects near and far, large and small, bright and dark; hence I do not see how it can enter the mind of anyone that I have simple-mindedly remained deceived in my observations.' The hundreds and thousands of experiments remind one of Hooke, and are most likely equally spurious. Cf. footnote 9 of Chapter 10.

25. Lagalla, *De phaenomenis in orbe lunae novi telescopii usa a D. Galileo Galilei nunc iterum suscitatis physica disputatio* (Venice, 1612), p. 8; quoted from E. Rosen, *The Naming of the Telescope*, New York, 1947, p. 54. The regular reports (*Avvisi*) of the Duchy of Urbino on events and gossip in Rome contain the following notice of the event: 'Galileo Galilei the mathematician, arrived here from Florence before Easter. Formerly a Professor at Padua, he is at present retained by the Grand Duke of Tuscany at a salary of 1,000 scudi. He has observed the motion of the stars with the *occiali*, which he invented or rather improved. Against the opinion of all ancient philosophers, he declares that there are four more stars or planets, which are satellites of Jupiter and which he calls the Medicean bodies, as well as two companions of Saturn. He has here discussed this opinion of his with father Clavius, the Jesuit. Thursday evening, at Monsignor Malavasia's estate outside the St Pancratius gate, a high and open place, a banquet was given for him by Frederick Cesi, the marquis of Monticelli and nephew of Cardinal Cesi, who was accompanied by his kinsman, Paul Monaldesco. In the gathering there were Galileo; a Fleming named Terrentius; Persio, of Cardinal Cesi's retinue; [La] Galla, Professor at the University here; the Greek, who is Cardinal Gonzaga's mathematician; Piffari, Professor at Siena; and as many as eight others. Some of them went out expressly to perform this observation, and even though they stayed until

Other reports confirm this and similar events. Galileo himself points to the 'number and importance of the benefits which the instrument may be expected to confer, when used by land or sea'.[26] The *terrestrial success* of the telescope was, therefore, assured. Its application to the *stars*, however, was an entirely different matter.

---

one o'clock in the morning, they still did not reach an agreement in their views' (quoted from Rosen, op. cit., p. 31).

26. *Sidereal Messenger*, op. cit., p. ii. According to Berellus (*De Vero Telescopii Inventore*, Hague, 1655, p. 4), Prince Moritz immediately realized the military value of the telescope and ordered that its invention – which Berellus attributes to Zacharias Jansen – be kept a secret. Thus the telescope seems to have commenced as a secret weapon and was turned to astronomical use only later. There are many anticipations of the telescope to be found in the literature, but they mostly belong to the domain of *natural magic* and are used accordingly. An example is Agrippa von Nettesheim, who, in his book on occult philosophy (written 1509, Book II, chapter 23), writes 'et ego novi ex illis miranda conficere, et specula in quibus quis videre poterit quaecunque voluerit a longissima distantia'. 'So may the toy of one age come to be the precious treasure of another,' Henry Morley, *The Life of Cornelius Agrippa von Nettesheim*, Vol. II, p. 166.

Appendix 1

The variation of the magnitudes of planets occasionally played an important role in the development of planetary theory. According to Simplicio's *De Coelo*, II, 12, Aristotle noticed the phenomenon but did not revise his astronomy of concentric spheres. Hipparchus ordered the magnitudes of fixed stars on a numerical scale from 1 (brightest stars) to 6 (barely visible) and determined them from the visibility of stars at dawn (Zinner, *Entstehung und Ausbreitung der Kopernikanischen Lehre*, Erlangen, 1943, p. 30), and he inferred radial movement from the change of brightness in fixed stars (Plinius, *Hist. Nat.*, II, 24) and planets (II, 13). Ptolemy, *Syntaxis*, IX, 2, defines the task of planetary theory as showing that 'the apparent anomalies all come about with the help of circular movements of (constant angular velocity)' and he proceeds to dealing with the two anomalies of locomotion *without* ever mentioning brightness. He 'saves' the anomalies in the sense that he accounts for them in terms of circles run through with constant angular velocity and *not* in the sense that he finds an *arbitrary* formula for predicting phenomena (that this sense of 'saving' is the correct one had been argued by F. Krafft, *Beiträge zur Geschichte der Wissenschaft und Technik*, No. 5, Wiesbaden, 1955, pp. 5ff). According to Simplicio, *De Coelo*, II, 12, and Proklos, *Hypotyposis*, I, 18, the phenomena *to be* 'saved' in this sense *include* the fact that 'the planets themselves change their brightness' and this change *is* saved 'by excenters and epicycles' (*Hypot.*, VII, 13). Later on, when the epicyclic machinery was regarded as a mere artifice of calculation (cf. Duhem, *To Save the Phenomena*, Chicago, 1969, for references), the change of brightness was removed from the domain of the phenomena to be saved and was occasionally even used as an argument against a literal interpretation of the change of distance between earth and planet (see below, on Osiander). However, some astronomers used the discrepancy between the variation of distance as calculated from some version of Ptolemy and the actual changes in magnitude as an argument against the system of epicycles. Examples are Henry of Hesse, *De improbatione concentricorum et epicyclorum* (1364), and Magister Julmann, *Tractatus de reprobationibus epicyclorum et eccentricorum* (1377) (paraphrase after Zinner, pp. 81ff). According to Henry von Hesse the brightness of Mars, as calculated after al-Farghani, varies at a ratio of about 1:100, while the comparison with a candle that is first placed at a distance where it looks like Mars at its brightest and then removed to ten times the distance shows that it should be invisible at its minimum. Magister Julmann calculates the changes of magnitude as 42:1 in the case of Venus, 11:1 in the case of Mars, 4:1 in the case of the moon and 3:1 for Jupiter, all of which, he says, contradicts observation. Regiomontanus refers to improbable changes of brightness in Venus and Mars (Zinner, p. 133).

Using the data of *Syntaxis*, X, 7, a calculation in the case of Mars gives a variation of diameter of about 1:8, a variation of disk of about 1:64 (which, in accordance with Euclid's optics is regarded as the correct measure for the variation of brightness). The actual variation is four magnitudes, i.e. between 1:16 and 1:28, i.e. *between one and four magnitudes different from the calculated magnitudes* (the spread is due to the spread in the basis of the magnitudes). In the case of Venus the difference is even more noticeable. Copernicus, *De Revol.*, Chapter 10, last paragraph, and Rheticus, *Narratio Prima* in E. Rosen (ed.), *Three Copernican Treatises*, New York, 1969, p. 137, regard the problem as solved – but it is not. In the *Commentariolus* the values for Mars are as follows: radius of the 'great circle', 23; radius of deferent, 38; radius of first epicycle, 5 (cf. Rosen, op. cit., pp. 74, 77), hence longest distance/shortest distance $\sim 50 + (38 - 25) + 5/(38 - 25) - 5 \sim 8$ as before (Galileo, op. cit., pp. 321f, gives the value 1:8 for Mars and 1:6 for Venus): if the estimates of magnitude available in the 14th to 17th centuries were sufficiently precise to discover a discrepancy between the Ptolemaic predictions and the actual variations – and Heinrich von Hesse, Regiomontanus, and Copernicus thought they were – *then the problem of planetary magnitudes occurs unchanged in* Copernicus (this is also the opinion of Derek Price, 'Contra Copernicus', loc. cit., 213).

The situation is recognized by the much-maligned Osiander who mentions the problem in his introduction to the *De Revol.*, turning it into an argument for the 'hypothetical', i.e. instrumentalistic nature of the Copernican cosmology. He writes: 'It is not necessary that these hypotheses be true; they need not even be like the truth; it suffices when they lead to calculations which agree with the results of observation; *except* someone should be so ignorant in matters of geometry and of optics that he is prepared to regard the epicycle of Venus as being like the truth and to assume that it is the cause of its being now forty (or more) degrees ahead of the sun, now the same behind it. *For who does not see that this assumption necessarily implies that the diameter of the planet when close to the earth must be four times as large than when it is at the point most remote from the earth, and its body more than sixty times as large – a fact which is contradicted by the experience of all ages.*' (My italics.)

The italicized passage which is suppressed both by the critics and by the friends of Osiander (Duhem, p. 66, quotes Osiander before and after the passage, but omits the passage itself) explains the nature of his instrumentalism. We know that he was an instrumentalist for philosophical as well as for tactical reasons (letter to Rheticus of 20 April 1541, reprinted in K. H. Burmeister, *Georg Joachim Rheticus*, III, Wiesbaden, 1968, p. 25) and because instrumentalism agreed with a powerful tradition in astronomy (letter to Copernicus of 20 April 1541, translated in Duhem, p. 68). *Now* we see that he had also physical reasons for his philosophy: Copernicus, interpreted realistically, was inconsistent with obvious facts. This point is not mentioned in Popper's bombastic 'Three Views Concerning Human Knowledge', *Conjectures and Refutations*, New York, 1962, pp. 97ff, where Osiander *is* quoted – but only up to the 'except' that introduces the physical reasons for his move. Popper's Osiander thus appears as a philosophical dogmatist while in actual fact he is a true Popperian: he takes refutations seriously. Cf. also my essay 'Realism and Instrumentalism' in *The Critical Approach*, ed. Bunge, New York, 1964. Osiander's argument is discussed and decisively rejected by Bruno, *La Cena de le Ceneri*, *Opere Italiane*, I, ed. Gentile, Bari, 1907, p. 64: 'The apparent magnitude of a radiating object does not permit us to infer its actual magnitude or its distance.' This is true, but it is not accepted by Galileo who needs the difficulty to improve his propaganda for the telescope.

Appendix 2

Machamer's essay, though designed to turn Galileo into a regular guy, methodologywise, does not invalidate my main argument, which is: Galileo violates important rules of scientific method which were invented by Aristotle, improved by Grosseteste (among others), canonized by the logical positivists (such as Carnap and Popper); Galileo succeeds, because he did not follow these rules; his contemporaries with very few exceptions overlooked fundamental difficulties that existed at the time; and modern science developed quickly, and in the 'right' direction (from the point of view of today's science-lovers) because of this negligence. *Ignorance was bliss.* Conversely, a more determined application of the canons of scientific method, a more determined search for relevant facts, a more critical attitude, far from accelerating this development, would have brought it to a standstill. These are the points I want to establish in my case study of Galileo. Keeping them in mind, what can we say about the arguments of Machamer and his associates?

'In discussing a point,' writes Machamer, 'Feyerabend continually . . . ignores other relevant passages'; by which he means that I only discuss what I think to be Galileo's bad points and neglect the many marvellous arguments he allegedly has for the motion of the earth. Considering my purpose, I could safely do so: in order to show that 'all ravens are black' is upheld by questionable means, it is sufficient to produce one white raven and to reveal the attempts at concealing it, turning it into a black raven, or bullying people into believing that it is really black; one may safely ignore the many black ravens which no doubt also exist. In order to show that 'the earth moves' is upheld by questionable means it is sufficient to produce a single difficulty for this view and to reveal all the attempts at concealing it or turning it into supporting evidence; one may again safely ignore the good points of the hypothesis which, incidentally, are much more fragile and ambiguous in the case of Galileo than in the case of the ravens: the *phases of Venus* which Machamer mentions, do not make terrestrial motion more plausible, as he sees himself (Tycho!), and Galileo also draws them wrongly, thus adding to the evidence *against* his point of view. The *theory of the tides*, which Machamer introduces in a prominent place as a major argument for the motion of the earth, can assume this function only if one disregards its difficulties (which were big enough to be known even to the most bleary-eyed sailor) in exactly the same way in which Galileo chooses to disregard the evidence against the motion of the earth (this is admitted by Machamer, p. 9). The fact – if it is a fact – that some minor lights among Galileo's contemporaries found it interesting, took it up, and worked on it only proves my point, viz. that research always violates major methodological rules and cannot proceed otherwise. The *greater coherence* of the Copernican system, p. 12, is an especially bad example for the author and an especially good example for me: in the *Commentariolus* Copernicus had indeed worked out a system that was simple, and more coherent than the Ptolemaic system. By the time he published the *De Revolutionibus* the greater simplicity and coherence had disappeared before the demand for an accurate representation of the planetary motions. Galileo disregards this loss of coherence and simplicity, for he disregards all epicycles. He returns to a theory even more primitive than the theory of the *Commentariolus* which is empirically inferior to Ptolemy. I do not criticize him for that (and for his silence on the problem of planetary motion). Quite the contrary – I think that was the only way in which progress could be made. In order to progress, we must step back from the evidence, reduce the degree of empirical adequacy (the empirical content) of our theories, abandon what we have already achieved, and start afresh. Almost all contemporary methodologists, Machamer included, think otherwise – and that is the point I wish to make.

To sum up this part of the debate: considering my aim, I could safely omit the 'arguments' which Galileo offered for the motion of the earth. *Adding* these arguments to the debate strengthens my case.

This is the place to make some minor methodological remarks. First, Machamer frequently misunderstands my way of arguing. Thus he objects to my saying that Kepler's optics is refuted by simple facts, because I have also stated that theories cannot be refuted by facts. This were a valid point if at the passage in question I had been talking to *myself*. Had I done so, then I would indeed have been forced to reply: 'But, my dear PKF, don't you remember that you have said that theories cannot be refuted by even the biggest fact?' But I did not talk to myself. I addressed people who *accept* the rule of falsification, and for *them* the example means trouble. Logicians are apt to call this an *argumentum ad hominem*. Quite so: in my essay I am addressing *humans*. I am addressing neither dogs nor logicians. Similar remarks apply to many other of Machamer's comments. (Incidentally, I would never accept the 'charitable' reading Machamer puts on my words in footnote 13. My argument is much more efficient as it stands.)

Secondly, Machamer often raises the ghosts of papers I wrote centuries ago (subjective time!) to combat something I wrote more recently. In this he is no doubt influenced by philosophers who, having made some tiny discovery, come back to it again and again for want of anything new to say and who turn this failing – lack of ideas – into a supreme virtue, viz. consistency. When writing a paper I have usually forgotten what I wrote before and application of earlier arguments is done at the applier's own risk.

Thirdly, Machamer misunderstands even those ideas which I still hold. I never said, as he assumes I did, that *any two* rival theories are incommensurable (footnote 35). What I *did* say was that *certain* rival theories, so-called 'universal' theories, or 'non-instantial' theories, *if interpreted in a certain way*, could not be compared easily. More specifically, I never assumed that Ptolemy and Copernicus are incommensurable. They are not.

Back to history. Machamer tries to show that the story of the telescope went very different from the way in which I tell it. To see who is right and who is wrong, let me repeat what I claim to be the case. My claim is twofold. (1) The optical theories existing at the time were not sufficient as a theoretical foundation for building the telescope, and part of these theories made its reliability doubtful after it had been invented; (2) Galileo did not know the optical theories of his time.

As regards (2) Machamer points out with a great show of scholarship that Galileo knew that light goes in straight lines and is reflected at equal angles, and that he also knew the basics of triangulations (this is what his references on pp. 14 and 15 amount to). *Sancta simplicitas!* Next time I say in a lecture on differential equations that Strawson and his stooges don't know mathematics somebody is going to get up and tell me that Strawson surely knows the multiplication table! To apply: when saying that Galileo did not know optics I did not mean to imply that he did not know baby-optics. What I meant was that he was ignorant of those parts of optics which *at the time in question* were *necessary for building the telescope*, assuming the telescope was built as a result of an insight into the basic principles of optics. What were these principles?

There were two elements of early 17th-century optics which were necessary, but not sufficient, for understanding the telescope. Neither of them was worked out in great detail, nor had they ever been combined into one coherent body of theory. They were (a) a knowledge of the *images* created by lenses, and (b) a knowledge of the things *seen through* a lens.

The first element is pure physics. Nowhere in the optical literature to which Machamer refers is there any account of the images projected by a convex lens. Pinhole images *without* lenses were difficult enough to explain (cf. the contortions Pecham goes through in his *Perspectiva*, *John Pecham and the Science of Optics*, ed. David Lindberg, Madison and London, 1970, pp. 67ff). The correct explanation (without lenses) is given by Maurolyocus, but it is only in 1611, one year after the *Sidereus Nuncius*, that his book appears in print. As regards the second element which seems to be unknown to Machamer the situation is much less reassuring. Pecham, who is aware of the constancy phenomenon (Lindberg, op. cit., p. 147), emphasizes that 'it is impossible to certify the size of an object seen under refracted rays' (p. 217), which means that for him the physiological optics of refractive media is deficient at a most important point: it does not tell us what the 'faculty of size' will do with refracted rays. Add to this the (Aristotelian) principle that perception, applied under extraordinary circumstances, gives results not in agreement with reality, and the difficulties of (a) and (b) when *separated* will be apparent.

In the telescope the two processes are combined to give a single effect. *Theoretically* there was no way to achieve the combination except on the basis of entirely new principles. These principles – false ones – were provided by Kepler in 1604 and 1611.

So far the historical situation. What has Machamer to say about it? Machamer writes: 'Anyone who had read Pecham . . . knew that any optical instrument constructed from lenses was explicable in terms of optical laws – the laws of refraction and the nature of light' (p. 18). We have seen that 'anyone who had read Pecham' would come to an entirely different conclusion. He would realize that the 'laws of refraction and the nature of light' do not suffice, that one has to consider the reactions of the eye and of the brain, and these reactions are unknown in the case of refracting media. He would realize that the reasoning needed to arrive at the telescope 'is crude enough to have been thought out by anyone having studied optics' (footnote 61) only if by 'optics' one means post-Keplerian optics: Machamer, who regards the laws of refraction as sufficient for understanding the telescope, who silently adopts the point of view of Kepler and projects it back to Pecham (who had argued against a simplified version of it), has no inkling of the achievement that lies in the transition from the older views to Kepler and Descartes. For while Kepler's (erroneous) ideas may seem crude to some 20th-century 'historians' of science who have swallowed them without examination, the *invention* of these ideas in the historical circumstances I have described was far from crude. Did Galileo make this remarkable invention? It seems *very* unlikely. No discussion is found in his letters or in his writings. In the schools the textbooks, such as Pecham, would be an upper limit of sophistication that was only rarely reached, and they were insufficient. Besides, they pointed in the wrong direction. It is of course possible that Galileo, ignoring the very detailed psychological laws enunciated in these books, used the law of refraction, took it for granted that greater angles mean greater size even in refracting media, and went ahead on that basis. I do not think it likely that he proceeded in this way, but if he did – and Machamer comes very close to suggesting that he did – then my case would again be strengthened: Galileo made progress by disregarding important facts (such as the constancy phenomenon), sensible solutions (which he either did not know or did not understand), and by pushing a false hypothesis (false even for Pecham, and with good reason) to the limit. Besides, Machamer's frequent references to the traditional textbooks would be quite irrelevant in this case.

Next comes the nature of Galileo's observations. I maintain that some of Galileo's telescopic observations were contradictory while others could be corrected by observations with the naked eye. As regards the last point, Machamer says that 'historically not one of Galileo's contemporaries presented this argument' (footnote 12). Incorrect and irrelevant. Kepler objected to the impression of smoothness of the moon's edge and invited Galileo to 'investigate the matter again'. And if no one else got into the act then this just shows that people didn't observe very carefully and were *therefore* ready to accept the new astronomical miracles of Galileo. Again ignorance, or sloppiness, was bliss. I am not at all impressed by Professor Righini's 'calculations' (p. 23), whatever they are. For such calculations you only need the *general* distribution of lights and shadows, which Galileo probably got right. Nor am I impressed by the fact that *some* people recognize *some* things in Galileo's moon. What I *am* impressed by is the great difference between Galileo's moon and what everyone could see with his naked eyes. If the difference is due to Galileo's attempt to *emphasize* certain aspects of the moon he thought essential, as Machamer suspects, then we are back at my thesis that Galileo often deviates from fact to make his point. So far Machamer's comments.

What Machamer does not mention are the paradoxical aspects of Galileo's observations, for example the fact that the moon looks rugged inside but perfectly smooth on the edges, or the fact that planets were enlarged while the fixed stars lost in size. Nobody but Kepler was bothered by such discrepancies, which again shows how little *thought* was devoted to the observations. (It was this *thoughtlessness* of his contemporaries which enabled Galileo to get ahead as well as he did.)

Machamer makes a big fuss (more than three pages) about ten lines which I devote to the difference between terrestrial and celestial observations. In these ten lines I say that there are physical reasons as well as psychological reasons for a difference. Machamer talks about the first, but not about the second. He states, quite correctly, that cosmological arguments were based on interplanetary triangulations from the very beginning and that even Aristotle assumed that light obeys the same laws in the heavens and on earth. Quite so, but this was not the point I tried to make. What I did try to say was that light, being an 'interdepartmental agency', had *special properties*, and that it was subjected to *different conditions* in both realms. A look at the history of the theories of light from Parmenides to Einstein confirms the first part of my assertion. The second part is much less conspicuous, nobody paid attention to it, and those who did pay attention on some occasions forgot it on others. Stars were regarded as condensation points in the celestial spheres (Aristotle, *De Coelo*, 289a11 and ff; Simplicio; many mediaeval authors), there was a change of material from air to fire to ether, yet nobody seemed to raise the problem of the refractions arising therefrom. Discussions started at the time of Tycho, in his exchange with Rothmann, and these discussions were duly commented upon by Kepler. Kepler even makes some assumption about the 'celestial essence' one of his reasons for not constructing the telescope. 'You,' he writes in his reply to Galileo's *Nuncius* (ed. Rosen, p. 18), 'putting aside all misgivings . . . turned directly to visual experimentation.' So it is quite true that opticians ignored the differences asserted by the cosmologists and boldly triangulated into space. In doing so they showed either gross negligence, or ignorance, or a complete disregard for the demands of consistency (which I do not support, but which are supported by even the meanest methodologist). Yet they were successful. Once more ignorance, or superficiality, or muddleheadedness turned out to be bliss. Machamer, who does not pay attention to the entire historical situation but only to the part that pleases him, is quite unaware of this fruitful disorderliness. So it is not at all surprising that he thinks he has found a historical mistake in my essay. (It should be added that Kepler argues about celestial essences despite Tycho's work on comets and on the Nova of 1572, and that Galileo defended the atmospheric nature of comets as late as 1630. This shows that 'the Aristotelian distinction' between a celestial realm and a terrestrial realm cannot have 'collapsed completely' by 1577 as Machamer insinuates (p. 21). It collapsed with some, it did not collapse with others, nor did it collapse without any trace. Here as elsewhere Machamer is quick to generalize from the attitude of those he finds congenial.) So much for the *physical* problems of celestial observations.

It is different with the *psychological* problems raised by the telescopic observations. These problems were seen by Pecham and others (such as Roger Bacon) and they still remain (moon illusion). At the time of Galileo they were tremendous, and they account for many strange reports (some of which are discussed in my text). The problems were comparable to the problems of somebody who, having never seen a lens before, looks for the first time into a very bad *microscope*. Not knowing what to expect (after all, one doesn't meet man-sized fleas on the sidewalk), he is unable to separate the properties of the 'object' from the 'illusions' created by the instrument (distortions; coloured fringes; discolouring; etc.) and he cannot make sense of the objects themselves. On the surface of the earth – with buildings, ships, etc. – the telescope will of course work well; these are familiar things and our knowledge of them eliminates most distortions just as our knowledge of a voice and a language eliminates the distortions of the telephone. The compensatory process does not work in the sky as the first observers soon noticed, *and said*. Thus it is true that the telescope causes illusions both in the sky and in the terrestrial cases (p. 20), but only the heavenly illusions were a real problem, for the reasons just stated. It is interesting to see that the combined effect of the physical difference and the psychological factor was realized by Pecham, who says that 'the sizes of stars are not known with complete veracity, since the sky is a more subtle body than air and fire' (op. cit., p. 219).

Machamer concludes his essay with the following warning. 'The history,' he says, 'must be done, and done well, before the philosophical implications can be considered' (p. 46). This is excellent advice – but why did he ignore it? I would also add that a little thinking must be done, and done well, before even the simplest historical fact can be considered.

10

Nor does the initial experience *with the telescope provide such reasons. The first telescopic observations of the sky are indistinct, indeterminate, contradictory and in conflict with what everyone can see with his unaided eyes. And the only theory that could have helped to separate telescopic illusions from veridical phenomena was refuted by simple tests.*

To start with, there is the problem of telescopic vision. This problem *is* different for celestial and terrestrial objects; and it was also *thought to be* different in the two cases.[1]

It was thought to be different because of the contemporary idea that celestial objects and terrestrial objects are formed from different materials and obey different laws. This idea entails that the result of an interaction of light (which connects both domains and has special properties) with terrestrial objects cannot, without further discussion, be extended to the sky. To this physical idea one added, entirely in accordance with the Aristotelian theory of knowledge[2] (and also with present views about the

1. This is hardly ever realized by those who argue (with Kästner, op. cit., p. 133) that 'one does not see how a telescope can be good and useful on the earth and yet deceive in the sky'. Kästner's comment is directed against Horky. See below, text to footnotes 9–16 of the present chapter.

2. For this theory cf. G. E. L. Owen, 'ΤΙΘΕΝΑΙ ΤΑ ΦΑΙΝΟΜΕΝΑ', *Aristote et les Problèmes de la methode*, Louvain, 1961, pp. 83–103. For the development of Aristotelian thought in the Middle Ages, cf. A. C. Crombie, *Robert Grosseteste and the Origins of Experimental Science*, Oxford, 1953, as well as Clemens Baumker, 'Witelo, ein Philosoph und Naturforscher des 13. Jahrhunderts', *Beiträge zur Geschichte der Philosophie des Mittelalters*, Bd. III, Münster, 1908. The relevant works of Aristotle are *Anal. Post.*, *De Anima*, *De Sensu*. Concerning the motion of the earth, cf. *De Coelo*, 293a28f: 'But there are many others who would agree that it is wrong to give the earth the central position, *looking for confirmation rather to theory, than to the facts of observation*' (my italics). As we saw in Chapter 7, this was precisely the manner in which Galileo introduced Copernicanism, *changing* experience so as to fit his favourite theory. That the

matter), the idea that the senses are *acquainted* with the close appearance of terrestrial objects and are, therefore, able to perceive them distinctly, even if the telescopic image should be vastly distorted, or disfigured by coloured fringes. The stars are not known from close by. Hence we cannot in their case use our *memory* for separating the contributions of the telescope and those which come from the object itself.[3] Moreover, all the familiar cues (such as background, overlap, knowledge of nearby size, etc.), which constitute and aid our vision on the surface of the earth, are absent when we are dealing with the sky, so that new and surprising phenomena are bound to occur.[4] Only a new theory of vision, containing both hypotheses concerning the behaviour of light within the telescope as well as hypotheses concerning the reaction of the eye under exceptional circumstances could have bridged the gulf between the heavens and the earth that was, and still is, such an obvious fact of physics and of astronomical observation.[5] We shall soon have occasion to comment on the theories that were available at the time and we shall

---

senses are acquainted with our everyday surroundings, but are liable to give misleading reports about objects outside this domain, is proved at once by the *appearance of the moon*. On the earth large but distant objects in familiar surroundings, such as mountains, are seen as being large, and far away. The appearance of the moon, however, gives us an entirely false idea of its distance and its size.

3. It is not too difficult to separate the letters of a familiar alphabet from a background of unfamiliar lines, even if they should happen to have been written with an almost illegible hand. No such separation is possible with letters which belong to an unfamiliar alphabet. The parts of such letters do not hang together to form distinct patterns which stand out from the background of general (optical) noise (in the manner described by K. Koffka, *Psychol. Bull.*, 19, 1922, pp. 551ff, partly reprinted in *Experiments in Visual Perception*, ed. M. D. Vernon, London, 1966; cf. also the article by Gottschaldt in the same volume).

4. For the importance of cues such as diaphragms, crossed wires, background, etc., in the localization and shape of the telescopic image and the strange situations arising when no cues are present cf. Chapter 4 of Ronchi, *Optics, The Science of Vision*, op. cit., especially pp. 151, 174, 189, 191, etc. Cf. also R. L. Gregory, *Eye and Brain*, New York, 1966, *passim* and p. 99 (on the autokinetic phenomenon). *Explorations in Transactional Psychology*, ed. F. P. Kilpatrick, New York, 1961, contains ample material on what happens in the absence of familiar cues.

5. It is for this reason that the 'deep study of the theory of refraction' which Galileo pretended to have carried out (text to footnote 15 of Chapter 9) would have been quite *insufficient* for establishing the usefulness of the telescope; cf. also footnote 16 of the present chapter.

see that they were unfit for the task and were refuted by plain and obvious facts. For the moment, I want to stay with the observations themselves and I want to comment on the contradictions and difficulties which arise when one tries to take the celestial results of the telescope at their face value, as indicating stable, objective properties of the things seen.

Some of these difficulties already announce themselves in a report of the contemporary *Avvisi*[6] which ends with the remark that 'even though they (the participants in the gathering described) went out expressly to perform this observation (of "four more stars or planets, which are satellites of Jupiter . . . as well as of two companions of Saturn"[7]), and even though they stayed until one in the morning, they still did not reach an agreement in their views.'

Another meeting that became notorious all over Europe makes the situation even clearer. About a year earlier, on 24 and 25 April 1610, Galileo took his telescope to the house of his opponent, Magini, in Bologna to demonstrate it to twenty-four professors of all faculties. Horky, Kepler's overly-excited pupil, wrote on this occasion;[8] 'I never slept on the 24th or 25th of April, day or night, but I tested the instrument of Galileo's in a thousand ways,[9] both on things here below and on those above. *Below it works wonderfully;* in the heavens it deceives one, as some fixed stars [Spica Virginis, for example, is mentioned, as well as a terrestrial flame] are seen double.[10] I have as witnesses most excellent men and noble doctors . . . and all have admitted the instrument to deceive. . . . This silenced Galileo and on the 26th he sadly left quite early in the morning . . . not even thanking Magini for his splendid meal. . . .' Magini wrote to Kepler on 26 May: 'He has achieved nothing, for more than twenty learned men were present; yet nobody has seen the new planets distinctly (nemo perfecte vidit); he will hardly be able to

6. Details in Chapter 9, footnote 25.

7. This is how the ring of Saturn was seen at the time. Cf. also R. L. Gregory, *The Intelligent Eye*, p. 119.

8. Galileo, *Opere*, Vol. X, p. 342 (my italics, referring to the difference, commented upon above, between celestial and terrestrial observations).

9. The 'hundreds' and 'thousands' of observations, trials, etc., which we find here again are hardly more than a rhetorical flourish (corresponding to our 'I have told you a thousand times'). They cannot be used to infer a life of incessant observation.

10. Here again we have a case where external clues are missing. Cf. Ronchi, op. cit. as regards the appearance of flames, small lights, etc.

keep them.'[11] A few months later (in a letter signed by Ruffini) he repeats: 'Only some with sharp vision were convinced to some extent.'[12] After these and other negative reports had reached Kepler from all sides, like a paper avalanche, he asked Galileo for witnesses:[13] 'I do not want to hide it from you that quite a few Italians have sent letters to Prague asserting that they could not see those stars [the moons of Jupiter] with your own telescope. I ask myself how it can be that so many deny the phenomenon, including those who use a telescope. Now, if I consider what occasionally happens to me, then I do not at all regard it as impossible that a single person may see what thousands are unable to see. . . .[14] Yet I regret that the confirmation by others should take so long in turning up. . . . Therefore, I beseech you, Galileo, give me witnesses as soon as possible. . . .' Galileo, in his reply of 19 August, refers to himself, to the Duke of Toscana, and Giuliano de Medici 'as well as many others in Pisa, Florence, Bologna, Venice and Padua, who, however, remain silent and hesitate. Most of them are entirely unable to distinguish Jupiter, or Mars, or even the Moon as a planet. . . .'[15] – not a very reassuring state of affairs, to say the least.

Today we understand a little better why the direct appeal to telescopic vision was bound to lead to disappointment, especially in the initial stages. The main reason, one already foreseen by Aristotle, was that the senses applied under abnormal conditions are liable to give an abnormal response. Some of the older historians had an inkling of the situation, but they speak *negatively*, they try to explain the *absence* of satisfactory observational reports, the *poverty* of what is seen in the telescope.[16]

11. Letter of 26 May, *Opere*, III.

12. ibid., p. 196.

13. Letter of 9 August 1610, quoted from Caspar-Dyck, *Johannes Kepler in Seinen Briefen*, Vol. I, Munich, 1930, p. 349.

14. Kepler, who suffered from Polyopia ('instead of a single small object at a great distance, two or three are seen by those who suffer from this defect. Hence, instead of a single moon ten or more present themselves to me', *Conversation*, op. cit., footnote 94; cf. also the remainder of the footnote for further quotations), and who was familiar with Platter's anatomical investigations (cf. S. L. Polyak, *The Retina*, Chicago, 1942, pp. 134ff for details and literature), was well aware of the need for a *physiological criticism of astronomical observations.*

15. Caspar-Dyck, *op. cit.*, p. 352.

16. Thus Emil Wohlwill, *Galileo und sein Kampf für die Kopernikanische Lehre*, Vol. I, Hamburg, 1909, p. 288, writes: 'No doubt the unpleasant results were due to

They are unaware of the possibility that the observers might have been disturbed by *strong positive illusions* also. The extent of such illusions was not realized until quite recently, mainly as the result of the work of Ronchi and his school.[17] Here the greatest variations are reported in the *placement* of the telescopic image and, correspondingly, in the observed *magnification*. Some observers put the image right inside the telescope making it change its lateral position with the lateral position of the eye, exactly as would be the case with an after image, or a reflex inside the telescope – an excellent proof that one must be dealing with an 'illusion'.[18] Others place the image in a manner that leads to no magnification at all, although a linear magnification of over thirty may have been promised.[19]

---

the lack of training in telescopic observation, and the restricted field of vision of the Galilean telescope as well as to the absence of any possibility for changing the distance of the glasses in order to make them fit the peculiarities of the eyes of the learned men. . . .' A similar judgement, though more dramatically expressed, is found in Arthur Köstler's *Sleepwalkers*, p. 369.

17. Cf. Ronchi, *Optics*, op. cit.: *Histoire de la Lumière*, Paris, 1956; *Storia del Cannochiale*, Vatican City, 1964; *Critica dei Fondamenti dell' Acustica e del'Ottica*, Rome, 1964; cf. also E. Cantore's summary in *Archives d'histoire des Sciences*, December 1966, pp. 333ff. I would like to acknowledge at this place that Professor Ronchi's investigations have greatly influenced my thinking on scientific method. For a brief historical account of Galileo's work cf. Ronchi's article in *Scientific Change*, ed. A. C. Crombie, London, 1963, pp. 542–61. How little this field is explored becomes clear from S. Tolansky's book *Optical Illusions*, London, 1964. Tolansky is a physicist who in his microscopic research (on crystals and metals) was distracted by one optical illusion after another. He writes: 'This turned our interest to the analysis of other situations, with the ultimate unexpected discovery that optical illusions can, and do, play a very real part in affecting many daily scientific observations. This warned me to be on the lookout and as a result I met more illusions than I had bargained for.' The 'illusions of direct vision', whose role in scientific research is slowly being rediscovered were well known to mediaeval writers on optics, who treated them in special chapters of their textbooks. Moreover, they treated lens-images as *psychological* phenomena, as results of a misapprehension, for an image 'is merely the appearance of an object outside its place' as we read in John Pecham (cf. David Lindberg, 'The "Perspectiva Communis" of John Pecham', *Archives Internationales d'histoire des sciences*, 1965, p. 51, as well as the last paragraph of Proposition ii/19 of Pecham's *Perspectiva Communis*, which is to be found in *John Pecham and the Science of Optics*, ed. D. Lindberg, Wisconsin, 1970, p. 171).

18. Ronchi, *Optics*, op. cit., p. 189. This may explain the frequently uttered desire to look *inside* the telescope. No such problems arise in the case of *terrestrial* objects whose images are regularly placed 'in the plane of the object' (ibid., p. 182).

19. For the magnification of Galileo's telescope cf. *The Sidereal Messenger*, op. cit.,

Even a doubling of images can be explained as the result of a lack of proper focusing.[20] Adding the many imperfections of the contemporary telescopes to these psychological difficulties,[21] one can well understand the scarcity of satisfactory reports and one is rather astonished at the speed with which the reality of the new phenomena was accepted, and, as was the custom, publicly acknowledged.[22] This development becomes

---

p. 11, cf. also A. Sonnefeld, 'Die Optischen Daten der Himmelsfernrohre von Galileo Galilei', *Jenaer Rundschau*, Vol. 7, 1962, pp. 207ff. The old rule 'that the size, position and arrangement according to which a thing is seen depends on the size of the angle through which it is seen' (R. Grosseteste, *De Iride*, quoted from Crombie, *Robert Grosseteste*, Oxford, 1953, p. 120), which goes back to Euclid, is *almost always wrong*. I still remember my disappointment when, having built a reflector with an alleged linear magnification of about 150, I found that the moon was only about five times enlarged, and situated quite close to the ocular (1937).

20. The image remains sharp and unchanged over a considerable interval – the lack of focusing may show itself in a doubling, however.

21. The first usable telescope which Kepler received from Elector Ernst of Köln (who in turn had received it from Galileo), and on which he based his *Narratio de observatis a se quartuor Jovis satellibus*, Frankfurt, 1611, showed the stars as *squares* and intensely *coloured* (*Ges. Werke,* IV, p. 461). Ernst von Köln himself was unable to see anything with the telescope and he asked Clavius to send him a better instrument (*Archivio della Pontifica Universita Gregoriana*, 530, f 182r). Francesco Fontana, who from 1643 onwards observed the phases of Venus, notes an unevenness of the boundary (and infers mountains), cf. R. Wolf, *Geschichte der Astronomie*, Munich, 1877, p. 398. For the idiosyncrasies of contemporary telescopes and descriptive literature cf. Ernst Zinner, *Deutsche und Niederländische Astronomische Instrumente des 11. bis 18. Jahrhunderts*, Munich, 1956, pp. 216–21. Refer also to the author catalogue in the second part of the book.

22. Father Clavius (letter of 17 December 1610, *Opere*, X, p. 485), the astronomer of the powerful Jesuit Collegium Romanum, praises Galileo as the first to have observed the moons of Jupiter and he recognizes their reality. Magini, Grienberger, and others soon followed suit. It is clear that, in doing so, they did not proceed according to the methods prescribed by their own philosophy, or else they were very lax in the investigation of the matter. Professor McMullin (op. cit., footnote 32) makes much of this quick acceptance of Galileo's telescopic observations: 'The regular periods observed for the satellites and for the phases of Venus strongly indicated that they were not artefacts of physiology or optics. There was surely no need for "auxiliary sciences"....' 'There was no need for auxiliary sciences,' writes McMullin, while using himself the unexamined auxiliary hypothesis that astronomical events are distinguished from physiological events by their regularity and their intersubjectivity. But this hypothesis is *false*, as is shown by the moon illusion, the phenomenon of fata morgana, the rainbow, haloes, by the many microscopic illusions which are so vividly described by Tolansky, by the phenomena of witchcraft (*every* woman reported an incubus to have an ice-cold member), and by numerous other phenomena. The hypothesis was also *known to be*

even more puzzling when we consider that many reports of even the best observers were either plainly *false*, and capable of being shown as such at the time, or else *self-contradictory*.

Thus Galileo reports unevennesses, 'vast protuberances, deep chasms, and sinuosities'[23] at the inner boundary of the lighted part of the moon while the outer boundary 'appear(s) not uneven, rugged, and irregular, but perfectly round and circular, as sharply defined as if marked out with a pair of compasses, and without the indentations of any protuberances and cavities'.[24] The moon, then, seemed to be full of mountains at the

---

false by Pecham, Witelo, and other mediaeval scholars who had studied the regular and intersubjective 'illusions' created by lenses, mirrors, and other optical contrivances. In antiquity the falsehood of McMullin's hypothesis was *commonplace*. Galileo explicitly discusses and repudiates it in his book on comets. Thus a new theory of vision was needed, not just to *accept* the Galilean observations, but also to provide *arguments* for their astronomical reality. Of course, Clavius may not have been aware of this need. This is hardly surprising. After all, some of his sophisticated 20th-century successors, such as Professor McMullin, are not aware of it either. In addition we must point out that the 'regular periods' of the moons of Jupiter were not as well known as McMullin insinuates. For his whole life Galileo tried to determine these periods in order to find better ways of determining longitude at sea. He did not succeed. Later on the same problem returned in a different form when the attempt to determine the velocity of light with more than one moon led to inconsistent results (Cassini). For the attitude of Clavius and the scientists of the Collegium Romanum cf. the very interesting book *Galileo in China* by Pasquale M. d'Elia, S.J., Harvard University Press, 1960. The early observations of the astronomers of the Collegium are contained in their own 'Nuncius Sidereus', *Ed. Naz.*, III/1, pp. 291–98.

23. *The Sidereal Messenger*, op. cit., p. 8.

24. op. cit., p. 24. – cf. the drawing on page 131 which is taken from Galileo's publication. Kepler in his *Optics* of 1604 writes (on the basis of observations with the unaided eye): 'It seemed as though something was missing in the circularity of the outmost periphery' (*Werke*, Vol. II, p. 219). He returns to this assertion in his *Conversation* (op. cit., pp. 28ff), criticizing Galileo's telescopic results by what he himself had seen with the unaided eye: 'You ask why the moon's outermost circle does not also appear irregular. I do not know how carefully you have thought about this subject or whether your query, as is more likely, is based on popular impression. For in my book [the *Optics* of 1604] I state that there was surely some imperfection in that outermost circle during full moon. Study the matter, and once again tell us, how it looks to you. . . .' Here the results of naked eye observation are quoted against Galileo's telescopic reports – and with perfectly good reason, as we shall see below. The reader who remembers Kepler's polyopia (cf. footnote 14 to this chapter) may wonder how he could trust his senses to such an extent. The reply is contained in the following quotation (*Werke*, II, pp. 194ff): 'When eclipses of the moon begin, I, who suffer from this defect, become aware of the eclipse before all the other observers. Long before the

inside but perfectly smooth at the periphery, and this despite the fact that the periphery *changed* as the result of the slight librations of the lunar body. The moon and some of the planets, such as for example Jupiter, were enlarged while the apparent diameter of the fixed stars decreased: the former were brought nearer whereas the latter were pushed away. 'The stars,' writes Galileo, 'fixed as well as erratic, when seen with the telescope, by no means appear to be increased in magnitude in the same proportion as other objects, and the Moon itself, gain increase of size; but in the case of the stars such increase appears much less, so that you may consider that a telescope, which (for the sake of illustration) is powerful enough to magnify other objects a hundred times, will scarcely render the stars magnified four or five times.'[25]

---

eclipse starts, I even detect the direction from which the shadow is approaching, while the others, who have very acute vision, are still in doubt. . . . The afore-mentioned waviness of the moon (cf. the previous quotation) stops for me when the moon approaches the shadow, and the strongest part of the sun's rays is cut off. . . .' Galileo has two explanations for the contradictory appearance of the moon. The one involves a lunar atmosphere (*Messenger*, op. cit., pp. 26ff). The other explanation (ibid., pp. 25ff), which involves the tangential appearance of series of mountains lying behind each other, is not really very plausible as the distribution of mountains near the visible side of the lunar globe does not show the arrangement that would be needed (this is now even better established by the publication of the Russian moon photograph of 7 October 1959; cf. Zdenek Kopal, *An Introduction to the Study of the Moon*, North Holland, 1966, p. 242).

25. *Messenger*, op. cit., p. 38; cf. also the more detailed account in *Dialogue*, op. cit., pp. 336ff. 'The telescope, as it were, removes the heavens from us,' writes A. Chwalina in his edition of *Kleomedes, Die Kreisbewegung der Gestirne* (Leipzig, 1927, p. 90), commenting on the decrease of the apparent diameter of *all* stars with the sole exception of the sun and the moon. Later on, the different magnification of planets (or comets) and fixed stars was used as a means of distinguishing them. 'From experience, I know,' writes Herschel in the paper reporting his first observation of Uranus (*Phil. Trans., 71*, 1781, pp. 493ff – the planet is here identified as a *comet*), 'that the diameters of the fixed stars are not proportionally magnified with higher powers, as the planets are; therefore, I now put on the powers of 460 and 932, and found the diameter of the comet increased in proportion to the power, as it ought to be. . . .' It is noteworthy that the rule did not invariably apply to the telescopes in use at Galileo's time. Thus, commenting on a comet of November 1618, Horatio Grassi ('On the Three Comets of 1618' in *The Controversy of the Comets of 1618*, op. cit., p. 17) points out 'that when the comet was observed through a telescope it suffered scarcely any enlargement', and he infers, perfectly in accordance with Herschel's 'experience', that 'it will have to be said that it is more remote from us than the moon. . . .' In his *Astronomical Balance* (ibid., p. 80) he repeats that, according to the common experience of 'illustrious astronomers' from

The strangest features of the early history of the telescope emerge, however, when we take a closer look at Galileo's *pictures of the moon*.

It needs only a brief look at Galileo's drawings, and at photographs of similar phases, to convince the reader that 'none of the features recorded . . . can be safely identified with any known markings of the lunar landscape'.[26] Looking at such evidence it is very easy to think that 'Galileo was not a great astronomical observer; or else that the excitement of so many telescopic discoveries made by him at that time had temporarily blurred his skill or critical sense'.[27]

Now this assertion may well be true (though I rather doubt it in view of the quite extraordinary observational skill which Galileo exhibits on other occasions).[28] But it is poor in content and, I submit, not very interesting. No new suggestions emerge for additional research, and the possibility of a *test* is rather remote.[29] There are, however, other hypotheses which do lead to new suggestions and which show us how complex the situation was at the time of Galileo. Let us consider the following two.

Hypothesis I. Galileo recorded faithfully what he saw and in this way left us evidence of the shortcomings of the first telescopes as well as of the peculiarities of contemporary telescopic vision. Interpreted in this

---

'many parts of Europe', 'the comet observed with a very extended telescope received scarcely any increment. . . .' Galileo (ibid., p. 177) accepts this as a fact, criticizing only the conclusions which Grassi wants to draw from it. All these phenomena refute Galileo's assertion (*Assayer*, op. cit., p. 204) that the telescope 'works always in the same way'. They also undermine his theory of irradiation (cf. footnote 55 to this chapter).

26. Kopal, op. cit., p. 207.

27. R. Wolf (*Geschichte der Astronomie*, p. 396) remarks on the poor quality of Galileo's drawings of the moon ('. . . seine Abbildung des Mondes kann man . . . kaum . . . eine Karte nennen'), while Zinner (*Geschichte der Sternkunde*, Berlin, 1931, p. 473) calls Galileo's observations of the moon and Venus 'typical for the observations of a beginner'. His picture of the moon, according to Zinner, 'has no similarity with the moon' (ibid., p. 472). Zinner also mentions the much better quality of the almost simultaneous observations made by the Jesuits (ibid., p. 473), and he finally asks whether Galileo's observations of the moon and Venus were not the result of a fertile brain, rather than of a careful eye ('sollte dabei . . . der Wunsch der Vater der Beobachtung gewesen sein?') – a very just question, especially in view of the phenomena briefly described in footnote 33 to this chapter.

28. The discovery and identification of the moons of Jupiter were no mean achievements, especially as a useful stable support for the telescope had not yet been developed.

29. The reason, among other things, is the great variation of telescopic vision from one observer to the next, cf. Ronchi, op cit., Chapter IV.

way Galileo's drawings are reports of exactly the same kind as are the reports emerging from the experiments of Stratton, Ehrismann, and Kohler[30] – except that the characteristics of the physical apparatus and the unfamiliarity of the objects seen must be taken into account too.[31] We must also remember the many conflicting views which were held about the surface of the moon, even at Galileo's time,[32] and which may have influenced what observers saw.[33] What would be needed in order to shed more light on the matter is an empirical collection of all the early telescopic results, preferably in parallel columns, including whatever pictorial representations have survived.[34] Subtracting instrumental peculiarities, such a collection adds fascinating material to a yet-to-be-written history of perception (and of science).[35] This is the content of Hypothesis I.

30. For a survey and some introductory literature cf. Gregory, op. cit., Chapter 11. For a more detailed discussion and literature cf. K. W. Smith and W. M. Smith, *Perception and Motion*, Philadelphia, 1962, reprinted in part in M. D. Vernon, op. cit. The reader should also consult Ames' article 'Aniseikonic Glasses', *Explorations in Transactional Psychology*, which deals with the change of *normal* vision caused by sometimes quite slightly abnormal optical conditions. A comprehensive account is given by I. Rock, *The Nature of Perceptual Adaptation*, New York, 1966.

31. Many of the old instruments, and excellent descriptions of them, are still available cf. Zinner, *Deutsche und Niederlandische astronomische Instrumente*.

32. For interesting information the reader should consult the relevant passages of Kepler's *Conversation* as well as of his *Somnium* (the latter is now available in a new translation by E. Rosen, who has added a considerable amount of background material: *Kepler's Somnium*, ed. Rosen, Madison, 1967). The standard work for the beliefs of the time is still Plutarch's *Face on the Moon* (it will be quoted from H. Cherniss' translation of *Moralia XII*, London, 1967).

33. 'One describes the moon after objects one thinks one can perceive on its surface' (Kästner, op. cit., Vol. IV, p. 167, commenting on Fontana's observational reports of 1646). 'Maestlin even saw rain on the moon' (Kepler, *Conversation*, op. cit., pp. 29f, presenting Maestlin's own observational report); cf. also da Vinci, notebooks, quoted from J. P. Richter, *The Notebooks of Leonardo da Vinci*, Vol. II, New York, 1970, p. 167: 'If you keep the details of the spots of the moon under observation you will often find great variation in them, and this I myself have proved by drawing them. And this is caused by the clouds that rise from the waters in the moon. . . .' For the instability of the images of unknown objects and their dependence on belief (or 'knowledge') cf. Ronchi, op. cit., Chapter 4.

34. Chapter 15 of Kopal, op. cit., contains an interesting collection of exactly this kind. Wider scope has W. Schulz, *Die Anschauung vom Monde und seinen Gestalten in Mythos und Kunst der Völker*, Berlin, 1912.

35. One must, of course, also investigate the dependence of what is seen on the

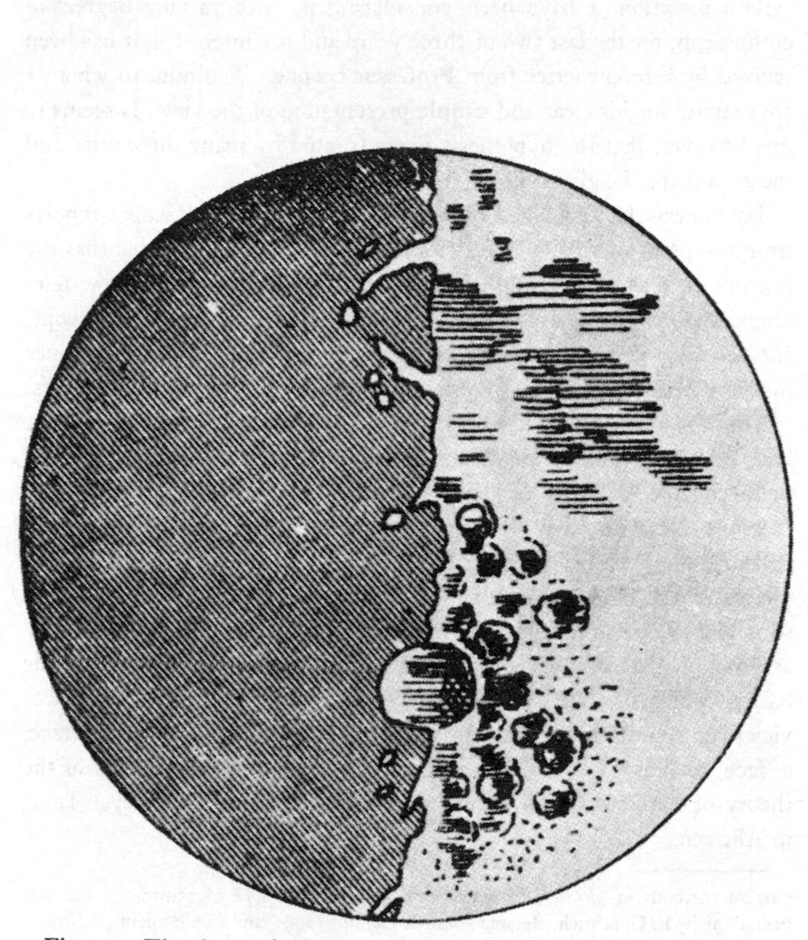

Figure 1. The shape of a lunar mountain and a walled plain, from Galileo, *Sidereus Nuncius*, Venice, 1610 (cf. p. 150).

Hypothesis II is more specific than Hypothesis I, and develops it in a certain direction. I have been considering it, with varying degrees of enthusiasm, for the last two or three years and my interest in it has been revived by a recent letter from Professor Stephen Toulmin, to whom I am grateful for his clear and simple presentation of the view. It seems to me, however, that the hypothesis is confronted by many difficulties and must, perhaps, be given up.

Hypothesis II, just like Hypothesis I, approaches telescopic reports from the point of view of the theory of perception; but it adds that the practice of telescopic observation and acquaintance with the new telescopic reports changed not only what was seen through the telescope, *but also what was seen with the naked eye*. It is obviously of importance for our evaluation of the contemporary attitude towards Galileo's reports.

That the appearance of the stars, and of the moon, may at some time have been much more indefinite than it is today was originally suggested to me by the existence of various theories about the moon which are incompatible with what everyone can plainly see with his own eyes. Anaximander's theory of partial stoppage (which aimed to explain the phases of the moon), Xenophanes' belief in the existence of different suns and different moons for different zones of the earth, Heraclitus' assumption that eclipses and phases are caused by the turning of the basins, which for him represented the sun and the moon[36] – all these views run counter to the existence of a stable and plainly visible surface, a 'face' such as we 'know' the moon to possess. The same is true of the theory of Berossos which occurs as late as Lucretius[37] and, even later, in Alhazen.

---

current methods of *pictorial representation*. Outside the field of astronomy this has been done by E. Gombrich, *Art and Illusion*, London, 1960, and L. Choulant, *A History and Bibliography of Anatomical Illustration*, New York, 1945 (translated, with additions, by Singer and others), who deals with anatomy. Astronomy has the advantage that *one* side of the puzzle, viz. the stars, is fairly simple in structure (much simpler than the uterus, for example) and relatively well known; cf. also Chapter 17 below.

36. For these theories and further literature cf. J. L. D. Dreyer, *A History of Astronomy from Thales to Kepler*, New York, 1953.

37. For Berossos, cf. Toulmin's article in *Isis*, No. 38, 1967, p. 65. Lucretius writes (*On the Nature of Things*, transl. Leonard, New York, 1957, p. 216): 'Again, she may revolve upon herself / like to a ball's sphere – if perchance that be – / one half of her dyed o'er with glowing light / and by the revolution of that sphere / she may beget for

Now such disregard for phenomena, which for us are quite obvious, may be due either to a certain indifference towards the existing evidence, which was, however, as clear and as detailed as it is today, *or else to a difference in the evidence itself*. It is not easy to choose between these alternatives. Having been influenced by Wittgenstein, Hanson, and others, I was for some time inclined towards the second version, but it now seems to me that it is ruled out both by physiology (psychology)[38] and by historical information. We need only remember how Copernicus disregarded the difficulties arising from the variations in the brightness of Mars and Venus, which were well known at the time.[39] And as regards the face of the moon, we see that Aristotle refers to it quite clearly when observing that 'the stars do not *roll*. For rolling involves rotation: but the "face", as it is called, of the moon is always seen.'[40] We may infer, then, that the occasional disregard for the stability of the face was due not to a lack of clear impressions, but to some widely held views about the unreliability of the senses. This inference is supported by Plutarch's discussion of the matter which plainly deals not with what is *seen* (except as evidence for or against certain views) but with certain *explanations* of phenomena otherwise *assumed to be well known*:[41] 'To begin with,' he says, 'it is absurd to call the figure seen in the moon an affection of vision . . . a condition which we call bedazzlement (glare). Anyone who asserts this does not observe that this phenomenon should rather have occurred in relation to the sun, since the sun lights upon us keen and

---

us her varying shapes / until she turns that fiery part of her / full to the sight and open eyes of men. . . .'

38. Cf. text to footnotes 50ff of my 'Reply to Criticism', op. cit., p. 246.

39. In antiquity the differences in the magnitudes of Venus and Mars were regarded as being 'obvious to our eyes', Simplicius, *De Coelo*, II, 12, Heiberg, p. 504. Polemarchus here considers the difficulties of Eudoxos' theory of homocentric spheres, viz. that Venus and Mars 'appear in the midst of the retrograde movement many times brighter, so that (Venus) on moonless nights causes bodies to throw shadows' (objection of Autolycus) and he may well be appealing to the possibility of a deception of the senses (which was frequently discussed by ancient schools). Aristotle, who must have been familiar with all these facts, does not mention them anywhere in *De Coelo* or in the *Metaphysics*, though he gives an account of Eudoxos' system and of the improvements of Polemarchus and Kalippus. Cf. footnote 7 of Chapter 9.

40. *De Coelo*, 290a25ff.

41. op cit., p. 37, cf. also S. Sambursky, *The Physical World of the Greeks*, New York, 1962, pp. 244ff.

violent, and moreover does not explain why dull and weak eyes discern no distinction of shape in the moon but her orb for them has an even and full light whereas those of keen and robust vision make out more precisely and distinctly the pattern of facial features and more clearly perceive the variations.' 'The unevenness also entirely refutes the hypothesis,' Plutarch continues,[42] 'for the shadow that one sees is not continuous and confused, but is not badly depictured by the words of Agesianax: "She gleams with fire encircled, but within / Bluer than lapis show a maiden's eye / And dainty brow, a visage manifest." In truth, the dark patches submerge beneath the bright ones which they encompass . . . and they are thoroughly entwined with each other so as to make the delineation of the figure resemble a painting.' Later on the stability of the face is used as an argument against theories which regard the moon as being made of fire, or air, for 'air is tenuous and without configuration, and so it naturally slips and does not stay in place'.[43] The *appearance* of the moon, then, seemed to be a well-known and distinct phenomenon. What was in question was the *relevance* of the phenomenon for astronomical theory.[44]

42. ibid., cf. however, footnote 17 to this chapter, Pliny's remark (*Hist. Nat.*, II, 43, 46) that the moon is 'now spotted and then suddenly shining clear', as well as da Vinci's report, referred to in footnote 33 to this chapter. 43. ibid., p. 50.

44. All this requires further research, especially in view of the contemporary distrust in vision as expressed in the principle *Non potest fieri scientia per visum solum*. Ronchi ('Complexities, Advances, and Misconceptions in the Development of the Science of Vision: What is being Discovered?', *Scientific Change*, op. cit., p. 544 – but note the criticism in D. C. Lindberg and N. H. Steneck, 'The Sense of Vision and the Origins of Modern Science' in *Science, Medicine and Society in the Renaissance*, New York, 1900) writes about this principle as follows: 'No scientific value should be attached to anything observed by sight alone. Visual observation could never be considered valid unless confirmation was available by touch.' As a consequence 'no one used the . . . enlarged images [created by concave mirrors] as the basis of a microscope. The reason for this essential fact is clear: nobody believed what he saw in a mirror, once he realized that he could not confirm it by touch.' There are also the surprising changes of normal terrestrial perception which can perhaps be inferred from the results of Snell and Dodds, cf. Chapter 17. It may also be a little unreasonable to assume that phenomena will be unaffected by one's views about their relation to the world. (After-images may be bright and disturbing for someone who has just obtained his sight. Later on they become almost unnoticeable and must be studied by special methods.) The hypothesis in the text is developed in one particular direction not so much because I am convinced that it is true but in order to indicate possible avenues of research and in order to give a clear impression of the complexity of the situation at Galileo's time.

We can safely assume that the same was true at the time of Galileo.[45]

But then we must admit that Galileo's observations could be checked with the naked eye and could in this way be exposed as illusory.

Thus the circular monster below the centre of the disk of the moon[46] is well above the threshold of naked eye observation (its diameter is larger than 3½ minutes of arc), while a single glance convinces us that the face of the moon is not anywhere disfigured by a blemish of this kind. It would be interesting to see what contemporary observers had to say on the matter[47] or, if they were artists, what they had to draw on the matter.

I summarize what has emerged so far.

Galileo was only slightly acquainted with contemporary optical *theory*.

45. A strong argument *in favour* of this contention is Kepler's description of the moon in his *Optics* of 1604: he comments on the broken character of the boundary between light and shadow (*Werke*, II, p. 218) and describes the dark part of the moon during an eclipse as looking like torn flesh or broken wood (ibid., p. 219). He returns to these passages in the *Conversation* (op. cit., p. 27), where he tells Galileo that 'these very acute observations of yours do not lack the support of even my own testimony. For [in my] *Optics* you have the half moon divided by a wavy line. From this fact I deduced peaks and depressions in the body of the moon. [Later on] I describe the moon during an eclipse as looking like torn flesh or broken wood, with bright streaks penetrating into the region of the shadow.' Remember also that Kepler criticizes Galileo's telescopic reports on the basis of his own naked-eye observations; cf. footnote 24 of this chapter.

46. 'There is one other point which I must on no account forget, which I have noticed and rather wondered at it. It is this: The middle of the Moon, as it seems, is occupied by a certain cavity larger than all the rest, and in shape perfectly round. I have looked at this depression near both the first and the third quarters, and I have represented it as well as I can in the second illustration already given. It produces the same appearance as to effects of light and shade as a tract like Bohemia would produce on the Earth, if it were shut in on all sides by very lofty mountains arranged on the circumference of a perfect circle; for the tract in the moon is walled in with peaks of such enormous height that the furthest side adjacent to the dark portion of the moon is seen bathed in sunlight before the boundary between light and shade reaches half way across the circular space . . . .' (*Messenger*, op. cit., pp. 21ff). This description, I think, definitely refutes Kopal's conjecture of observational laxity. It is interesting to note the difference between the woodcuts in the *Nuncius* (p. 131, Figure I) and Galileo's original drawing. The woodcut corresponds quite closely to the description while the original drawing with its impressionistic features ('Kaum eine Karte,' says Wolf) is vague enough to escape the accusation of gross observational error.

47. 'I cannot help wondering about the meaning of that large circular cavity in what I usually call the left corner of the mouth,' writes Kepler (*Conversation*, op. cit., p. 28), and then proceeds to make conjectures as to its origin (conscious efforts by intelligent beings included).

His telescope gave surprising results on the earth, and these results were duly praised. Trouble was to be expected in the sky, as we know now. Trouble promptly arose: the telescope produced spurious and contradictory phenomena and some of its results could be refuted by a simple look with the unaided eye. Only a new *theory* of telescopic vision could possibly bring order into the chaos (which may have been still larger, due to the different phenomena seen at the time even with the naked eye) and could separate appearance from reality. Such a theory was developed by Kepler, first in 1604 and then again in 1611.[48]

According to Kepler, the place of the image of a punctiform object is found by first tracing the path of the rays emerging from the object according to the laws of (reflection and) refraction until they reach the eye, and by then using the principle (still taught today) that 'the image will be seen in the point determined by the backward intersection of the rays of vision from both eyes'[49] or, in the case of monocular vision, from the two sides of the pupil.[50] This rule, which proceeds from the assumption that 'the image is the work of the act of vision', is partly empirical and partly geometrical.[51] It bases the position of the image on a 'metrical triangle'[52] or a 'telemetric triangle', as Ronchi calls it,[53] that is constructed out of the rays which finally arrive at the eye and is used by the eye *and the mind* to place the image at the proper distance. Whatever the optical system, whatever the total path of the rays from the object

48. I have here disregarded the work of della Porta (*De Refractione*) and of Maurolycus who both anticipated Kepler in certain respects (and are duly mentioned by him). Maurolycus makes the important step [*Photismi de Lumine*, transl. Henry Crew, New York, 1940, p. 45 (on mirrors) and p. 74 (on lenses)] of considering only the cusp of the caustic; but a connection with what is seen on *direct* vision is still not established. For the difficulties which were removed by Kepler's simple and ingenious hypothesis cf. Ronchi, *Histoire de la Lumière*, op. cit., Chapter III.

49. *Werke*, II, p. 72. The *Optics* of 1604 has been partly translated into German by F. Plehn, *J. Keplers Grundlagen der geometrischen Optik*, Leipzig, 1922. The relevant passages occur in section 2 of Chapter 3, pp. 38–48.

50. ibid., p. 67.

51. 'Cum imago sit visus opus', ibid., p. 64. 'In visione tenet sensus communis oculorum suorum distantiam ex assuefactione, angulos vero ad illam distantiam notat ex sensu contortionis oculorum', ibid., p. 66.

52. 'Triangulum distantiae mensorium', ibid., p. 67.

53. *Optics, the Science of Vision*, op. cit., p. 44. One should also consult the second chapter of this book for the history of pre-Keplerian optics.

to the observer, the mind of the observer utilizes its *very last part only* and bases its visual judgement, the perception, on it.

It is clear that this rule constituted a considerable advance over and above all previous thought. However, it needs only a second to show that it is entirely false: take a magnifying glass, determine its focus, and look at an object close to it. The telemetric triangle now reaches beyond the object to infinity. A slight change of distance brings the Keplerian image from infinity to close by and back to infinity. No such phenomenon is ever observed. We see the image, slightly enlarged, in a distance that is most of the time identical with the actual distance between the object and the lens. The visual distance of the image remains constant, however much we may vary the distance between lens and object and even when the image becomes distorted and, finally, diffuse.[54]

54. Ronchi, *Optics*, pp. 182, 202. This phenomenon was known to everyone who had used a magnifying glass only once, Kepler included. Which shows that disregard of familiar phenomena does not entail that the phenomena were seen differently (cf. text to footnote 44 to this chapter). Isaac Barrow's account of the difficulty of Kepler's rule was mentioned above (text to footnote 16 to Chapter 5). According to Berkeley (op. cit., p. 141) 'this phenomenon . . . entirely subverts the opinion of those who will have us judge of distances by lines and angles. . . .' Berkeley replaces this opinion by his own theory according to which the mind judges distances from the clarity or confusion of the primary impressions. Kepler's idea of the telemetric triangle was adopted at once by almost all thinkers in the field. It was given a fundamental position by Descartes according to whom 'Distantiam . . . discimus, per mutuam quandam conspirationem oculorum' (*Dioptrice*, quoted from *Renati Descartes Specima Philosophiae*, Amsterdam, 1657, p. 87). 'But,' says Barrow, 'neither this nor any other difficulty shall . . . make me renounce that which I know to be manifestly agreeable to reason.' It is this attitude which was responsible for the slow advance of a scientific theory of *eye glasses* and of optics in general. 'The reason for this peculiar phenomenon,' writes Moritz von Rohr (*Das Brillenglas als optisches Instrument*, Berlin, 1934, p. 1), 'is to be sought in the close connection between the eye glass and the eye and it is impossible to give an acceptable theory of eye glasses without understanding what happens in the process of vision itself. . . .' The telemetric triangle omits precisely this process, or rather gives a simplistic and false account of it. The state of optics at the beginning of the 20th century is well described in A. Gullstrand's 'Appendices to Part I' of Helmholtz' *Treatise on Physiological Optics*, transl. Southall, New York, 1962, pp. 261ff. We read here how a return to the psycho-physiological process of vision enabled physicists to arrive at a more reasonable account even of the physics of optical imagery: 'The reason why the laws of actual optical imagery have been, so to speak, summoned to life by the requirements of physiological optics is due partly to the fact that by means of trigonometrical calculations, tedious to be sure, but easy to perform, it has been possible for the optical engineer to get closer to the realities of his problem. Thus, thanks to the

This, then, was the actual situation in 1610 when Galileo published his telescopic findings. How did Galileo react to it? The answer has already been given: he raised the telescope to the state of a 'superior and better sense'.[55] What were his reasons for doing so? This question

---

labours of such men as Abbe and his school, technical optics has attained its present splendid development; whereas, with the scientific means available, a comprehensive grasp of the intricate relations in the case of the imagery in the eye has been actually impossible.'

55. 'O Nicholas Copernicus, what a pleasure it would have been for you to see this part of your system confirmed by so clear an experiment!' writes Galileo, implying that the new telescopic phenomena are additional support for Copernicus (*Dialogue*, op. cit., p. 339). The difference in the appearance of planets and fixed stars (cf. footnote 27 to this chapter) he explains by the hypothesis that 'the very instrument of seeing [the eye] introduces a hindrance of its own' (ibid., p. 335), and that the telescope removes this hindrance, viz. *irradiation*, permitting the eye to see the stars and the planets as they really are. (Mario Giuducci, a follower of Galileo, ascribed irradiation to refraction by moisture on the surface of the eye, *Discourse on the Comets of 1618*, op. cit., p. 47.) This explanation, plausible as it may seem (especially in view of Galileo's attempt to show how irradiation can be removed by means other than the telescope) is not as straightforward as one might wish. Gullstrand (op. cit., p. 426) says that 'owing to the properties of the wave surface of the bundle of rays refracted in the eye . . . it is a mathematical impossibility for any cross section to cut the caustic surface in a smooth curve in the form of a circle concentric with the pupil'. Other authors point to 'inhomogeneities in the various humours, and above all in the crystalline lens' (Ronchi, *Optics*, op. cit., p. 104). Kepler gives this account (*Conversation*, op. cit., pp. 33ff): 'Point sources of light transmit their cones to the crystalline lens. There refraction takes place, and behind the lens the cones again contract to a point. But this point does not reach as far as the retina. Therefore, the light is dispersed once more, and spreads over a small area of the retina, whereas it should impinge on a point. Hence the telescope, by introducing another refraction, makes this point coincide with the retina. . . .' Polyak, in his classical work *The Retina*, attributes irradiation partly to 'defects of the dioptrical media and to the imperfect accommodation' but 'chiefly' to the 'peculiar structural constitution of the retina itself' (p. 176), adding that it may be a function of the brain also (p. 429). None of these hypotheses covers *all* the facts known about irradiation. Gullstrand, Ronchi, and Polyak (if we omit his reference to the brain which can be made to explain anything we want) cannot explain the disappearance of irradiation in the telescope. Kepler, Gullstrand and Ronchi also fail to give an account of the fact, emphasized by Ronchi, that large objects show no irradiation at their edges ('Anyone undertaking to account for the phenomenon of irradiation must admit that when he looks at an electric bulb from afar so that it seems like a point, he sees it surrounded by an immense crown of rays whereas from nearby he sees nothing at all around it,' *Optics*, op. cit., p. 105). We know now that large objects are made definite by the lateral inhibitory interaction of retinal elements (which is further increased by brain function), cf. Ratliff, *Mach Bands*, p. 146, but the variation of the phenomenon

brings me back to the problems raised by the evidence (against Copernicus) that was reported and discussed in Chapter 9.

---

with the diameter of the object and under the conditions of telescopic vision remains unexplored. Galileo's hypothesis received support mainly from its agreement with the Copernican point of view and was, therefore, largely *ad hoc*.

11

On the other hand, there are some telescopic phenomena which are plainly Copernican. Galileo introduces these phenomena as independent evidence for Copernicus while the situation is rather that one refuted view – Copernicanism – has a certain similarity to phenomena emerging from another refuted view – the idea that telescopic phenomena are faithful images of the sky. Galileo prevails because of his style and his clever techniques of persuasion, because he writes in Italian rather than in Latin, and because he appeals to people who are temperamentally opposed to the old ideas and the standards of learning connected with them.

According to the Copernican theory, Mars and Venus approach and recede from the earth by a factor of 1:6 or 1:8, respectively. (These are approximate numbers.) Their change of brightness should be 1:40 and 1:60, respectively (these are Galileo's values). Yet Mars changes very little and the variation in the brightness of Venus 'is almost imperceptible'.[1] These experiences 'overtly contradict the annual movement [of the earth]'.[2] The telescope, on the other hand, produces new and strange *phenomena*, some of them exposable as illusory by observation with the naked eye, some contradictory, some having even the appearance of being illusory, while the only *theory* that could have brought order into this chaos, Kepler's theory of vision, is refuted by evidence of the plainest kind possible. But – and with this I come to what I think is the central feature of Galileo's procedure – *there are telescopic phenomena*, namely the telescopic variation of the brightness of the planets, *which agree more closely with Copernicus than do the results of naked-eye observation.*

1. The actual variations of Mars and Venus are four magnitudes and one magnitude respectively.
2. *Dialogue*, op. cit., p. 328.

Seen through the telescope, Mars does indeed change as it should according to the Copernican view. Compared with the total performance of the telescope this change is still quite puzzling. It is just as puzzling as is the Copernican theory when compared with the pre-telescopic evidence. But the change is in harmony with the predictions of Copernicus. *It is this harmony* rather than any deep understanding of cosmology and of optics *which for Galileo proves Copernicus and the veracity of the telescope* in terrestrial *as well as* celestial matters. And it is this harmony on which he builds an entirely new view of the universe. 'Galileo,' writes Ludovico Geymonat,[3] referring to this aspect of the situation, 'was not the first to turn the telescope upon the heavens, but... he was the first to grasp the enormous interest of the things thus seen. And he understood at once that these things fitted in perfectly with the Copernican theory whereas they contradicted the old astronomy. Galileo had believed for years in the truth of Copernicanism, but he had never been able to demonstrate it, despite his exceedingly optimistic statements to friends and colleagues [he had not even been able to remove the refuting instances, as we have seen, and as he says himself]. Should direct proof [should even mere *agreement* with the evidence] be at last sought here? The more this conviction took root in his mind, the clearer to him became the importance of the new instrument. In Galileo's own mind faith in the reliability of the telescope and recognition of its importance were not *two separate acts*, rather, they were *two aspects of the same process*.' Can the absence of independent evidence be expressed more clearly? 'The *Nuncius*,' writes Franz Hammer in the most concise account I have read of the matter,[4] 'contains two unknowns, the one being solved with the help of the other.' This is entirely correct, except that the 'unknowns' were not so much unknown as known to be false,

3. op. cit., pp. 38ff (my italics).

4. *Johannes Kepler, Gesammelte Werke*, op. cit., Vol. IV, p. 447. Kepler (*Conversation*, op. cit., p. 14) speaks of 'mutually self-supporting evidence'. Remember, however, that what is 'mutually self-supporting' are two *refuted* hypotheses (or two hypotheses which may even be *incommensurable* with the available basic statements) and *not* two hypotheses which have *independent support* in the domain of basic statements. In a letter to Herwarth of 26 March 1598, Kepler speaks of the 'many reasons' he wants to adduce for the motion of the earth, adding that 'each of these reasons, taken for itself, would find only scant belief' (Caspar-Dyck, *Johannes Kepler in seinen Briefen*, Vol. I, Munich, 1930, p. 68).

as Galileo says himself. It is this rather peculiar situation, this harmony between two interesting but refuted ideas which Galileo exploits in order to prevent the elimination of either.

Exactly the same procedure is used to preserve his new dynamics. We have seen that this science, too, was endangered by observable events. To eliminate the danger Galileo introduces friction and other disturbances with the help of *ad hoc* hypotheses, treating them as tendencies *defined* by the obvious discrepancy between fact and theory rather than as physical events *explained* by a theory of friction for which new and independent evidence might some day become available (such a theory arose only much later, in the 18th century). Yet the agreement between the new dynamics and the idea of the motion of the earth, which Galileo increases with the help of his method of *anamnesis*, makes both seem more reasonable.

The reader will realize that a more detailed study of historical phenomena such as these, creates considerable difficulties for the view that the transition from the pre-Copernican cosmology to that of the 17th century consisted in the replacement of refuted theories by more general conjectures which explained the refuting instances, made new predictions, and were corroborated by the observations carried out to test these new predictions. And he will perhaps see the merits of a different view which asserts that, while the pre-Copernican astronomy *was in trouble* (was confronted by a series of refuting instances and implausibilities, the Copernican theory *was in even greater trouble* (was confronted by even more drastic refuting instances and implausibilities); but that being in harmony *with still further inadequate theories* it gained strength, and was retained, the refutations being made ineffective by *ad hoc* hypotheses and clever techniques of persuasion. This would seem to be a much more adequate description of the developments at the time of Galileo than is offered by almost all alternative accounts.

I shall now interrupt the historical narrative to show that the description is not only *factually adequate*, but that it is also *perfectly reasonable*, and that any attempt to enforce some of the more familiar methodologies of the 20th century – such as, for example, the method of conjectures and refutations – would have had disastrous consequences.

12

Such 'irrational' methods of support are needed because of the 'uneven development' (Marx, Lenin) of different parts of science. Copernicanism and other essential ingredients of modern science survived only because reason was frequently overruled in their past.

A prevalent tendency in methodological discussions is to approach problems of knowledge *sub specie aeternitatis*, as it were. Statements are compared with each other without regard to their history and without considering that they might belong to different historical strata. For example, one asks: given background knowledge, initial conditions, basic principles, accepted observations – what conclusions can we draw about a newly suggested hypothesis? The answers vary considerably. Some say that it is possible to determine degrees of confirmation and that the hypothesis can be evaluated with their help. Others reject any logic of confirmation and judge hypothesis by their content, and by the falsifications that have actually occurred. But almost everyone takes it for granted that precise observations, clear principles and well-confirmed theories *are already decisive*; that they can and must be used *here and now* to either eliminate the suggested hypothesis, or to make it acceptable, or perhaps even prove it![1]

1. In a series of interesting and provocative papers Professor Kurt Huebner of the University of Kiel has criticized the 'abstract' character of contemporary methodologies and he has maintained that 'the source of scientific progress lies neither in abstract rules of falsification, nor in inductive inferences and the like, but in the entire mental and historical situation in which a scientist finds himself. It is from *this* situation that he takes his presuppositions and it is upon it that his activity acts back. . . . The decisive weakness of contemporary philosophy of science seems to me to lie in this: despite the great variety of schools and thinkers it still proceeds unhistorically. It tries to solve its basic problems – the character of the methods to be applied and the justification of the statements obtained with their help – by mere reflection, where thinking apparently is only left to itself and to its sophistication . . .' ('Was zeigt Kepler's "Astronomia Nova"

Such a procedure makes sense only if we can assume that the elements of our knowledge – the theories, the observations, the principles of our arguments – are *timeless entities* which share the same degree of perfection, are all equally accessible, and are related to each other in a way that is independent of the events that produced them. This is, of course, an extremely common assumption. It is taken for granted by every logician; it underlies the familiar distinction between a context of discovery and a context of justification; and it is often expressed by saying that science deals with propositions and not with statements or sentences. However, the procedure overlooks that science is a complex and heterogeneous *historical process* which contains vague and incoherent anticipations of future ideologies side by side with highly sophisticated theoretical systems and ancient and petrified forms of thought. Some of its elements are available in the form of neatly written statements while others are submerged and become known only by contrast, by comparison with new and unusual views. (This is the way in which the inverted tower argument helped Galileo to discover the natural interpretations hostile to Copernicus. And this is also the way in which Einstein discovered certain deep-lying assumptions of classical mechanics, such as the assumption of the existence of infinitely fast signals. For general considerations, cf. the last paragraph of Chapter 5.) Many of the conflicts and contradictions which occur in science are due to this heterogeneity of the material, to this 'unevenness' of the historical development, as a Marxist would say, and they have no immediate theoretical significance.[2]

---

der modernen Wissenschaftstheorie?' in *Philosophia Naturalis*, Vol. 11, 1969, pp. 267ff). Huebner also examines the strange development that leads from historically oriented thinkers such as Duhem, Mach, Poincaré, Meyerson and others to the dry, unhistorical and, therefore, essentially unscientific attitude of today (*Phil. Nat.*, No. 13, 1971, pp. 81–97), and he is preparing a theory of science that takes history into account by giving a sketch of a 'Structural Theory of History' (*Studium Generale*, No. 24, 1971, pp. 851–64, especially pp. 858ff). This is the way that will have to be followed in the future if one wants to overcome the sterility of present-day philosophy of science.

2. According to Marx, 'secondary' parts of the social process, such as demand, artistic production or legal relations, may get ahead of material production and drag it along: cf. *The Poverty of Philosophy* but especially the *Introduction to the Critique of Political Economy*, Chicago, 1918, p. 309: 'The unequal relation between the development of material production and art, for instance. In general, the conception of progress

They have much in common with the problems which arise when a power station is needed right next to a Gothic cathedral. Occasionally, such features are taken into account; for example, when it is asserted that physical laws (statements) and biological laws (statements) belong to different conceptual domains and cannot be directly compared. But in most cases, and especially in the case observation vs. theory, our methodologies project all the various elements of science and the different historical strata they occupy on to one and the same plane, and proceed at once to render comparative judgements. This is like arranging a fight between an infant and a grown man, and announcing triumphantly, what is obvious anyway, that the man is going to win (the history of the kinetic theory and the more recent history of hidden variable theories in quantum mechanics is full of inane criticisms of this kind and so is the history of

---

is not to be taken in the sense of the usual abstraction. In the case of art, etc., it is not so important and difficult to understand this disproportion as in that of practical social relations, e.g. the relation between education in the U.S. and Europe. The really difficult point, however, that is to be discussed here is that of the unequal development of relations of production as legal relations.' Trotsky describes the same situation: 'The gist of the matter lies in this, that the different aspects of the historical progress – economics, politics, the state, the growth of the working class – do not develop simultaneously along parallel lines' ('The School of Revolutionary Strategy', speech delivered at the general party membership meeting of the Moscow Organization of July 1921, published in *The First Five Years of the Communist International*, Vol. II, New York, 1953, p. 5). See also Lenin, *Left-Wing Communism – an Infantile Disorder* (op. cit., p. 59), concerning the fact that multiple causes of an event may be out of phase and have an effect only when they occur together. In a different form, the thesis of 'uneven development' deals with the fact that capitalism has reached different stages in different countries, and even in different parts of the same country. This second type of uneven development may lead to inverse relations between the accompanying ideologies, so that efficiency in production and radical political ideas develop in inverse proportions. 'In civilized Europe, with its highly developed machine industry, its rich, multiform culture and its constitutions, a point of history has been reached when the commanding bourgeoisie, fearing the growth and increasing strength of the proletariat, comes out in support of everything backward, moribund, and medieval. . . . But all young Asia grows a mighty democratic movement, spreading and gaining in strength' (Lenin, 'Backward Europe and Advanced Asia', *Collected Works*, Vol. 19, op. cit., pp. 99ff). For this very interesting situation, which deserves to be exploited for the philosophy of science, cf. A. C. Meyer, *Leninism*, Chapter 12, Cambridge, 1957, and L. Althusser, *For Marx*, London and New York, 1970, Chapters 3 and 6. The philosophical background is splendidly explained in Mao Tse-tung's essay *On Contradiction* (*Selected Readings,* Peking, 1970, p. 70, especially section IV).

psychoanalysis and of Marxism). In our examination of new hypotheses we must obviously take the historical situation into account. Let us see how this is going to affect our judgement!

The geocentric hypothesis and Aristotle's theory of knowledge and perception are well adapted to each other. Perception supports the theory of locomotion that entails the unmoved earth and it is in turn a special case of a comprehensive view of motion that includes locomotion, increase and decrease, qualitative alteration, generation and corruption. This comprehensive view defines motion as the transition of a form from an agent to a patient which terminates when the patient possesses exactly the same form that characterized the agent at the beginning of the interaction. Perception, accordingly, is a process in which the form of the object perceived enters the percipient as precisely the same form that characterized the object so that the percipient, in a sense, assumes the properties of the object.

A theory of perception of this kind (which one might regard as a sophisticated version of naive realism) does not permit any major discrepancy between observations and the things observed. 'That there should be things in the world which are inaccessible to man not only now, and for the time being, but in principle, and because of his natural endowment, and which would therefore never be seen by him – this was quite inconceivable for later antiquity as well as for the Middle Ages.'[3] Nor does the theory encourage the use of instruments, for they interfere with the processes in the medium. These processes carry a true picture only as long as they are left undisturbed. Disturbances create forms which are no longer identical with the shape of the objects perceived – they create *illusions*. Such illusions can be readily demonstrated by examining

3. F. Blumenberg, *Galileo Galilei, Sidereus Nuncius, Nachricht von neuen Sternen*, Vol. I, Frankfurt, 1965, p. 13. Aristotle himself was more open-minded: 'The evidence (concerning celestial phenomena) is furnished but scantily by sensations, whereas respecting perishable plants and animals we have abundant information, living as we do in their midst . . .', *De Part. Anim.*, 644b26ff. In what follows, a highly idealized account is given of later Aristotelianism. Unless otherwise stated, the word 'Aristotle' refers to this idealization. For the difficulties in forming a coherent picture of Aristotle *himself* cf. Düring, *Aristoteles*, Heidelberg, 1966. For some differences between Aristotle and his mediaeval followers cf. Wolfgang Wieland, *Die Aristotelische Physik*, Göttingen, 1970.

the images produced by curved mirrors, or by crude lenses (and remember that the lenses used by Galileo were far from the level of perfection achieved today): they are distorted, the lens-images also have coloured fringes, and they may appear at a place different from the place of the object. Astronomy, physics, psychology, epistemology – all these disciplines collaborate in the Aristotelian philosophy to create a system that is coherent, rational and in agreement with the results of observation as can be seen from an examination of Aristotelian philosophy in the form in which it was developed by some mediaeval philosophers. Such an analysis shows the inherent power of the Aristotelian system).

The role of observation in Aristotle is quite interesting. Aristotle is an empiricist. His injunctions against an overly-theoretical approach are as militant as those of the 'scientific' empiricists of the 17th and 18th centuries. But while the latter take both the truth and the content of empiricism for granted, Aristotle explains (1) the nature of experience and (2) why it is important. Experience *is* what a normal observer (an observer whose senses are in good order and who is not drunk or sleepy, etc.) perceives under normal circumstances (broad daylight; no interference with the medium) and describes in an idiom that fits the facts and can be understood by all. Experience is *important for knowledge* because, given normal circumstances, the perceptions of the observer contain identically the same forms that reside in the object. Nor are these explanations *ad hoc*. They are a direct consequence of Aristotle's general theory of motion, taken in conjunction with the physiological idea that sensations obey the same physical laws as does the rest of the universe. And they are confirmed by the evidence that confirms either of these two views (the existence of distorted lens-images being part of the evidence). We understand today a little better why a theory of motion and perception which is now regarded as false could be so successful (evolutionary explanation of the adaptation of organisms; movement in media). The fact remains that no decisive empirical argument could be raised against it (though it was not free from difficulties).

This harmony between human perception and the Aristotelian cosmology is regarded as illusory by the supporters of the motion of the earth. In the view of the Copernicans there exist large-scale processes which involve vast cosmic masses and yet *leave no trace* in our experience.

Figure 2. Moon, age seven days (first quarter).

The existent observations therefore count no longer as tests of the new basic laws that are being proposed. They are not directly attached to these laws, and they may be entirely disconnected. *Today, after* the success of modern science has made us realize that the relation between man and the universe is not as simple as is assumed by naive realism, we can say that this was a correct guess, that the observer is indeed separated from the laws of the world by the special physical conditions of his observation platform, the moving earth (gravitational effects; law of inertia; Coriolis forces; influence of the atmosphere upon optical observations; aberration; stellar parallax; and so on . . .), by the idiosyncrasies of his basic instrument of observation, the human eye (irradiation; after-images; mutual inhibition of adjacent retinal elements; and so on . . .) as well as by older views which have invaded the observation language and made it speak the language of naive realism (natural interpretations). Observations may contain a contribution from the thing observed, but this contribution is usually overlaid by other effects (some of which we have just mentioned), and it may be completely obliterated by them. Just consider the image of a fixed star as viewed through a telescope. This image is displaced by the effects of refraction, aberration and, possibly, of gravitation. It contains the spectrum of the star not as it is now, but as it was some time ago (in the case of extragalactic supernovae the difference may be millions of years), and distorted by Doppler effect, intervening galactic matter, etc. Moreover, the extension and the internal structure of the image is entirely determined by the telescope and the eyes of the observer: it is the telescope that decides how large the diffraction disks are going to be, and it is the human eye that decides how much of the structure of these disks is going to be seen. One needs considerable skill *and much theory* to isolate the contribution of the original cause, the star, and to use it for a test, but this means that non-Aristotelian cosmologies can be tested only after we have *separated* observations and laws with the help of auxiliary sciences describing the complex processes that occur between the eye and the object, and the even more complex processes between the cornea and the brain. In the case of Copernicus we need a new *meteorology* (in the good old sense of the word, as dealing with things below the moon), a new science of *physiological optics* that deals with the subjective (mind) and the objective (light, medium, lenses, structure of the eye) aspects of vision as well as a new *dynamics* stating the manner in which the motion of the earth might influence the physical processes at its surface. Observations become relevant only *after* the processes described by these new subjects have been inserted between the world and the eye. The language in which we express our observations may have to be revised as well so that the new cosmology receives a fair chance and is not endangered by an unnoticed collaboration of sensations and older ideas. In sum: *what is needed for a test of Copernicus is an entirely new world view containing a new view of man and of his capacities of knowing*.

It is obvious that such a new world view will take a long time appearing, and that we may never succeed to formulate it in its entirety. It is extremely unlikely that the idea of the motion of the earth will at once be followed by the arrival, in full formal splendour, of all the sciences that are now said to constitute the body of 'classical physics'. Or, to be a little more realistic, such a sequence of events is not only extremely unlikely, *it is impossible in principle*, given the nature of man and the complexities of the world he inhabits. Yet it is only *after* these sciences have arrived that a test can be said to make sense.

This need to *wait*, and to *ignore* large masses of critical observations and measurements, is hardly ever discussed in our methodologies. Disregarding the possibility that a new physics or a new astronomy might have to be judged by a new theory of knowledge and might require entirely new tests, scientists at once confront it with the *status quo* and announce triumphantly that 'it is not in agreement with facts and received principles'. They are of course right, and even trivially so, but not in the sense intended by them. For at an early stage of development the contradiction only indicates that the old and the new are *different* and *out of phase*. It does not show which view is the *better* one. A judgement of *this* kind presupposes that the competitors confront each other on equal terms. How shall we proceed in order to bring about such a fair comparison?

The first step is clear: we must *retain* the new cosmology until it has been supplemented by the necessary auxiliary sciences. We must retain it in the face of plain and unambiguous refuting facts. We may, of course, try to explain our action by saying that the critical observations are either not relevant or that they are illusory, but we cannot support such an explanation by a single objective reason. Whatever explanation we give is nothing but a *verbal gesture*, a gentle invitation to participate in the development of the new philosophy. Nor can we reasonably remove the received *theory* of perception which says that the observations are relevant, gives reasons for this assertion, and is confirmed by independent evidence. Thus the new view is quite arbitrarily separated from those data that supported its predecessor and is made more 'metaphysical': a new period in the history of science commences with a *backward movement* that returns us to an earlier stage where theories were more vague and had smaller empirical content. This backward movement is not just an accident; it has a definite function; it is essential if we want to overtake the *status quo*, for it gives us the time and the freedom that are needed for developing the main view in detail, and for finding the necessary auxiliary sciences.[4]

This backward movement is indeed essential – but how can we persuade people to follow our lead? How can we lure them away from a well-defined, sophisticated and empirically successful system and make them transfer their allegiance to an unfinished and absurd hypothesis? To a hypothesis, moreover, that is contradicted by one observation after another if we only take the trouble to compare it with what is plainly shown to be the case by our senses? How can we convince them that the success of the *status quo* is only apparent and is bound to be shown as such in 500 years or more, when there is not a single argument on our side (and remember that the illustrations I used two paragraphs earlier derive their force from the successes of classical physics and were not available to the Copernicans).[5] It is clear that allegiance to the new ideas will have to be brought about by means other than arguments. It will

4. An example of a backward movement of this kind is Galileo's return to the kinematics of the *Commentariolus* and his disregard for the machinery of epicycles as developed in the *De Revol.* For an admirable *rational* account of this step cf. Imre Lakatos' talk 'A Philosopher looks at the Copernican Revolution', Leeds, 6 January 1973. (I have a typescript of the talk, kindly sent to me by Professor Lakatos.)

5. They were available to the sceptics, especially to Aenesidemus who points out, following Philo, that no object appears as it is but is modified by being combined with air, light, humidity, heat, etc.; cf. *Diogenes Laertius*, IX, 84. However, it seems that the sceptical view had only little influence on the development of modern astronomy, and understandably so: one does not start a movement by being reasonable.

have to be brought about *by irrational means* such as propaganda, emotion, *ad hoc* hypotheses, and appeal to prejudices of all kinds. We need these 'irrational means' in order to uphold what is nothing but a blind faith until we have found the auxiliary sciences, the facts, the arguments that turn the faith into sound 'knowledge'.

It is in this context that the rise of a new secular class with a new outlook and considerable contempt for the science of the schools, for its methods, its results, even for its language, becomes so important. The barbaric Latin spoken by the scholars (it has much in common with the no less barbaric 'ordinary English' spoken by Oxford philosophers), the intellectual squalor of academic science, its other-worldliness which is soon interpreted as uselessness, its connection with the Church – all these elements are now lumped together with the Aristotelian cosmology and the contempt one feels for them is transferred to every single Aristotelian argument.[6] This guilt-by-association does not make the arguments less *rational*, or less conclusive, *but it reduces their influence* on the minds of those who are willing to follow Copernicus. For Copernicus now stands for progress in other areas as well, he is a symbol for the ideals of a new class that looks back to the classical times of Plato and Cicero and forward to a free and pluralistic society. The association of astronomical ideas and historical and class tendencies does not produce new arguments either. But it engenders a firm commitment to the heliocentric view – and this is all that is needed at this stage, as we have seen. We have also seen how masterfully Galileo exploits the situation and how he amplifies it by tricks, jokes, and *non-sequiturs* of his own.

We are here dealing with a situation that must be analysed and understood if we want to adopt a more reasonable attitude towards the issue between 'reason' and 'irrationality' than is found in the school philosophies of today. Reason grants that the ideas which we introduce in order to expand and to improve our knowledge may *arise* in a very disorderly way and that the *origin* of a particular point of view may depend on class prejudice, passion, personal idiosyncrasies, questions of

6. For these social pressures cf. Olschki's magnificent *Geschichte der neusprachlichen wissenschaftlichen Literatur*. For the role of Puritanism cf. R. F. Jones, op. cit., Chapters V and VI.

style, and even on error, pure and simple. But it also demands that in *judging* such ideas we follow certain well-defined rules: our *evaluation* of ideas must not be invaded by irrational elements. Now, what our historical examples seem to show is this: there are situations when our most liberal judgements and our most liberal rules would have eliminated an idea or a point of view which we regard today as essential for science, and would not have permitted it to prevail – and such situations occur quite frequently (cf. for this point, the examples in Chapter 5). The ideas survived and they can *now* be said to be in agreement with reason. They survived because prejudice, passion, conceit, errors, sheer pig-headedness, in short because all the elements that characterize the context of discovery, *opposed* the dictates of reason *and because these irrational elements were permitted to have their way.* To express it differently: *Copernicanism and other 'rational' views exist today only because reason was overruled at some time in their past.* (The opposite is also true: witchcraft and other 'irrational' views have *ceased* to be influential only because reason was overruled at some time in *their* past.)[7]

Now, assuming that Copernicanism is a Good Thing, we must also admit that its survival is a Good Thing. And, considering the conditions of its survival, we must further admit that it was a Good Thing that reason was overruled in the 16th, 17th and even the 18th centuries. Moreover, the cosmologists of the 16th and 17th centuries did not have the knowledge we have today, they did not know that Copernicanism was capable of giving rise to a scientific system that is acceptable from the point of view of 'scientific method'. They did not know which of the many views that existed at their time would lead to future reason when defended in an 'irrational' way. Being without such guidance they had to make a guess and in making this guess they could only follow their inclinations, as we have seen. Hence it is advisable to let one's

7. These considerations refute J. Dorling who, in *British Journal for the Philosophy of Science*, Vol. 23, 1972, 189f, presents my 'irrationalism' as a presupposition of my research, not as a result. He continues: '. . . one would have thought that the philosopher of science would be most interested in picking out and analysing in detail those scientific arguments which did seem to be rationally reconstructible.' One would have thought that the philosopher of science would be most interested in picking out and analysing in detail those moves which are necessary for the *advancement* of science. Such moves, I have tried to show, resist rational reconstruction.

inclinations go against reason *in any circumstances*, for science may profit from it.[8]

It is clear that this argument, that advises us not to let reason overrule our inclinations and occasionally (or frequently – see again the material in Chapter 5) to suspend reason altogether, does not depend on the historical material which I have presented. If my account of Galileo is historically correct, then the argument stands as formulated. If it turns out to be a fairy-tale, then this fairy-tale tells us that a conflict between reason and the preconditions of progress is *possible*, it indicates how it might arise, and it forces us to conclude that our chances to progress *may* be obstructed by our desire to be rational. And note that progress is here defined as a rationalistic lover of science would define it, i.e. as entailing that Copernicus is better than Aristotle and Einstein better than Newton. Of course, there is no need to accept this definition which is certainly quite narrow. We use it only to show that an idea of reason accepted by the majority of rationalists (including all critical rationalists) may prevent progress as defined by the very same majority. I now resume the discussion of some details of the transition from Aristotle to Copernicus.

The first step on the way to a new cosmology, I have said, is a step *back*: apparently relevant evidence is pushed aside, new data are brought in by *ad hoc* connections, the empirical content of science is drastically reduced.[9] Now the cosmology that happens to be at the centre of attention and whose adoption causes us to carry out the changes just described differs from other views in one respect only: it has features which at the time in question seem attractive to some people. But there is hardly any idea that is totally without merit and that might not also become the starting point of concentrated effort. No invention is ever made in isolation, and no idea is, therefore, completely without (abstract or empirical) support. Now if partial support and partial plausibility suffice to start a new trend – and I have suggested that they do – if starting a new trend means taking a step back from the evidence, if any idea can

8. 'Reason' here includes the more liberalized rationality of our contemporary critical rationalists.

9. It is interesting to see that this is exactly what happens in the case of the quantum theory and of the theory of relativity. Cf. my essay 'Problems of Empiricism, Part II', *Pittsburgh Studies*, Vol. IV, Pittsburgh, 1970, sections 9 and 10.

become plausible and can receive partial support, then the step back is in fact a step forward, and away from the tyranny of tightly-knit, highly corroborated, and gracelessly presented theoretical systems. 'Another different error,' writes Bacon on precisely this point,[10] 'is the . . . peremptory reduction of knowledge into arts and methods, from which time the sciences are seldom improved; for as young men rarely grow in stature after their shape and limbs are fully formed, so knowledge, whilst it lies in aphorisms and observations, remains in a growing state; but when once fashioned into methods, though it may be further polished, illustrated and fitted for use, is no longer increased in bulk and substance.'

The similarity with the arts which has often been asserted arises at exactly this point. Once it has been realized that close empirical fit is no virtue and that it must be relaxed in times of change, then style, elegance of expression, simplicity of presentation, tension of plot and narrative, and seductiveness of content become important features of our knowledge. They give life to what is said and help us to overcome the resistance of the observational material.[11] They *create* and maintain interest in a theory that has been partly removed from the observational plane and would be inferior to its rivals when judged by the customary standards. It is in this context that much of Galileo's work should be seen. This work has often been likened to *propaganda*[12] – and propaganda it certainly is. But propaganda of this kind is not a marginal affair that may or may not be added to allegedly more substantial means of defence, and that should perhaps be avoided by the 'professionally honest scientist'. In the circumstances we are considering now, *propaganda is of the essence*. It is of the essence because interest must be created at a time when the usual methodological prescriptions have no point of attack; and because this interest must be maintained, perhaps for centuries, until new reasons arrive. It is also clear that such reasons, i.e. the appropriate auxiliary sciences, need not at once turn up in full formal splendour. They may at first be quite inarticulate, and may even conflict with the

10. *Advancement of Learning* (1605 edition), New York, 1944, p. 21. Cf. also the *Novum Organum*, Aphorisms 79, 86, as well as J. W. N. Watkins' splendid little book *Hobbes' System of Ideas*, London, 1965, p. 169.

11. 'What restitutes to scientific phenomenon its life, is art' (*The Diary of Anais Nin*, Vol. I, p. 277).

12. Cf. A. Koyré, *Etudes Galiléennes*, Vol. III, Paris, 1939, pp. 53ff.

existing evidence. Agreement, or partial agreement, with the cosmology is all that is needed in the beginning. The agreement shows that they are at least *relevant* and that they may some day produce full-fledged positive evidence. Thus the idea that the telescope shows the world as it really is leads to many difficulties. But the support it lends to, and receives from, Copernicus is a hint that we might be moving in the right direction.

We have here an extremely interesting relation between a general view and the particular hypotheses which constitute its evidence. It is often assumed that general views do not mean much unless the relevant evidence can be fully specified. Carnap, for example, asserts that 'there is no independent interpretation for [the language in terms of which a certain theory or world view is formulated]. The system *T* [the axioms of the theory and the rules of derivation] is itself an uninterpreted postulate system. [Its] terms obtain only an indirect and incomplete interpretation by the fact that some of them are connected by correspondence rules with observational terms.'[13] 'There is no independent interpretation,' says Carnap and yet an idea such as the idea of the motion of the earth, which is inconsistent (and perhaps even incommensurable) with contemporary evidence, which is upheld by declaring this evidence to be irrelevant and which is therefore cut off from the most important facts of contemporary astronomy, manages to become a nucleus, a crystallization point for the aggregation of other inadequate views which gradually increase in articulation and finally fuse into a new cosmology including new kinds of evidence. There is no better account of this process than the description which John Stuart Mill has left us of the vicissitudes of his education. Referring to the explanations which his father gave him on logical matters he writes: 'The explanations did not make the matter at all clear to me at the time; but they were not therefore useless; they remained as a nucleus for my observations and reflections to crystallize upon; the import of his general remarks being interpreted to me, by the particular instances which came under my notice *afterwards*.'[14] In exactly

13. 'The Methodological Character of Theoretical Concepts', *Minnesota Studies in the Philosophy of Science*, Vol. I, Minneapolis, p. 47.

14. *Autobiography*, quoted from *Essential Works of John Stuart Mill*, ed. Lerner New York, 1965, p. 21.

the same manner the Copernican view, though devoid of cognitive content from the point of view of a strict empiricism or else refuted, was needed in the construction of the supplementary sciences *even before* it became testable with their help and even before it, in turn, provided them with supporting evidence of the most forceful kind. Is it not clear that our beautiful and shining methodologies which demand from us that we concentrate on theories of high empirical content, which implore us to take risks and to take refutations seriously, and which compare statements that belong to different historical strata as if they were all equally perfect platonic ideas, would have given extremely bad advice in the circumstances? (The advice to *test* his theories would have been quite useless for Galileo, who was faced by an embarrassing amount of *prima facie* refuting instances, who was unable to *explain* them for he lacked the necessary knowledge [though not the necessary intuitions] and who had, therefore, to *explain them away*, so that a potentially valuable hypothesis might be saved from premature extinction.) And is it not also clear that we must become more realistic, that we must cease gaping at the imaginary shapes of an ideal philosophical heaven (a 'third world' as Popper now calls it), and must start considering what will help us in this *material* world, given our erring brains, our imperfect measuring instruments and our faulty theories? One can only marvel at how reluctant philosophers and scientists are to adapt their general views to an activity in which the latter already participate (and which, if asked, they would not want to give up). It is this reluctance, this psychological resistance which makes it necessary to combine abstract argument with the sledgehammer of history. Abstract argument is necessary, because it gives our thoughts *direction*. History, however, is necessary also, at least in the present state of philosophy, because it gives our arguments *force*. This explains my long excursion into 17th-century physics and astronomy.

To sum up the content of the last six chapters:

When the 'Pythagorean idea' of the motion of the earth was revived by Copernicus it met with difficulties which exceeded the difficulties encountered by contemporary Ptolemaic astronomy. Strictly speaking, one had to regard it as refuted. Galileo, who was convinced of the truth of the Copernican view and who did not share the quite common, though by no means universal, belief in a stable experience, looked for new kinds of fact which might support Copernicus and still be acceptable to all. Such facts he obtained in two different ways. First, by the invention of his *telescope* which changed the *sensory core* of everyday experience and replaced it by puzzling and unexplained phenomena; and by his *principle of relativity and his dynamics* which changed its *conceptual components*. Neither the telescopic phenomena nor the new ideas of motion were acceptable to common sense (or to the Aristotelians). Besides, the associated theories could be easily shown to be false. Yet these false theories, these unacceptable phenomena, are distorted by Galileo and are converted into strong support of Copernicus. The whole rich reservoir of the everyday experience and of the intuition of his readers is utilized in the argument, but the facts which they are invited to recall are arranged in a new way, approximations are made, known effects are omitted, different conceptual lines are drawn, so that *a new kind of experience* arises, *manufactured* almost out of thin air. This new experience is then *solidified* by insinuating that the reader has been familiar with it all the time. It is solidified and soon accepted as gospel truth, despite the fact that its conceptual components are incomparably more speculative than are the conceptual components of common sense. We may therefore say that Galileo's science rests on an *illustrated metaphysics*. The distortion permits Galileo to advance, but it prevents almost everyone else from making his effort the basis of a critical philosophy (even today, emphasis is put either on his mathematics, or on his alleged experiments, or on his frequent appeal to the 'truth', and his propagandistic moves are altogether neglected). I suggest that what Galileo did was to let refuted theories support each other, that he built in this way a new world-view which was only loosely (if at all!) connected with the preceding cosmology (everyday experience included), that he established fake connections with the perceptual elements of this cosmology which are only now being replaced by genuine theories (physiological optics, theory of continua), and that whenever possible he replaced old facts by a new type of experience which he simply *invented* for the purpose of supporting Copernicus. Remember, incidentally, that Galileo's procedure drastically reduces the content of dynamics: Aristotelian dynamics was a general theory of change comprising locomotion, qualitative change, generation and corruption. Galileo's dynamics and its successors deal with locomotion only, other kinds of motion being pushed aside with the promissory note (due to Democritos) that locomotion will eventually be capable of comprehending *all* motion. Thus, a comprehensive empirical theory of motion is replaced by a much narrower theory plus a metaphysics of motion, just as an 'empirical' experience is replaced by an experience that contains speculative elements. This, I suggest, was the actual procedure followed by Galileo. Proceeding in this way he exhibited a style, a sense of humour, an elasticity and elegance, and an awareness of the valuable weaknesses of human thinking, which has never been equalled in the history of science. Here is an almost inexhaustible source of material for methodological speculation and, much more importantly, for the recovery of those features of knowledge which not only inform, but which also delight us.

13

Galileo's method works in other fields as well. For example, it can be used to eliminate the existing arguments against materialism and to put an end to the philosophical *mind/body problem. (The corresponding* scientific *problems remain untouched, however.)*

Galileo made progress by changing familiar connections between words and words (he introduced new concepts), words and impressions (he introduced new natural interpretations), by using new and unfamiliar principles (such as his law of inertia and his principle of universal relativity), and by altering the sensory core of his observation statements. His motive was the wish to accommodate the Copernican point of view. Copernicanism clashes with some obvious facts, it is inconsistent with plausible, and apparently well-established, principles, and it does not fit in with the 'grammar' of a commonly spoken idiom. It does not fit in with the 'form of life' that contains these facts, principles, and grammatical rules. But neither the rules, nor the principles, nor even the facts are sacrosanct. The fault may lie with them and not with the idea that the earth moves. We may therefore change them, create new facts and new grammatical rules, and see what happens once these rules are available and have become familiar. Such an attempt may take considerable time, and in a sense the Galilean venture is not finished even today. But we can already see that the changes were wise ones to make and that it would have been foolish to stick with the Aristotelian form of life to the exclusion of everything else.

With the mind/body problem, the situation is exactly the same. We have again observations, concepts, general principles, and grammatical rules which, taken together, constitute a 'form of life' that apparently supports some views, such as dualism, and excludes others, such as materialism. (I say 'apparently' for the situation is much less clear here than it was in the astronomical case.) And we may again proceed in the Galilean manner, look for new natural interpretations, new facts, new grammatical rules, new principles which can accommodate materialism and then compare the *total* systems – materialism and the new facts, rules, natural interpretations, and principles on the one side; dualism and the old 'forms of life' on the other. Thus there is no need to try, like Smart, to show that materialism is compatible with the ideology of common sense. Nor is the suggested procedure as 'desperate' (Armstrong) as it must appear to those who are unfamiliar with conceptual change. The procedure was commonplace in antiquity and it occurs wherever imaginative researchers strike out in new directions (Einstein and Bohr are recent examples).[1]

1. For a more detailed discussion the reader is referred to Chapters 9–15 of my essay 'Problems of Empiricism', *Beyond the Edge of Certainty*, ed. Colodny, New York, 1965, preferably in the improved version published in Italian, *I problemi dell' Empirismo*. Milan, 1971, pp. 31–69.

14

The results obtained so far suggest abolishing the distinction between a context of discovery and a context of justification and disregarding the related distinction between observational terms and theoretical terms. Neither distinction plays a role in scientific practice. Attempts to enforce them would have disastrous consequences.

Let us now use the material of the preceding sections to throw light on the following features of contemporary empiricism: (1) the distinction between a context of discovery and a context of justification; (2) the distinction between observational terms and theoretical terms; (3) the problem of incommensurability. The last problem will lead us back to the problem of rationality and order vs. anarchism, which is the main topic of this essay.

One of the objections which may be raised against my attempt to draw methodological conclusions from historical examples is that it confounds two contexts which are essentially distinct, viz. a context of discovery, and a context of justification. *Discovery* may be irrational and need not follow any recognized method. *Justification*, on the other hand, or – to use the Holy Word of a different school – *criticism*, starts only *after* the discoveries have been made, and it proceeds in an orderly way. 'It is one thing,' writes Herbert Feigl, 'to retrace the historical origins, the psychological genesis and development, the socio-political-economic conditions for the acceptance or rejection of scientific theories; and it is quite another thing to provide a logical reconstruction of the conceptual structure and of the testing of scientific theories.'[1] These are indeed two different *things*, especially as they are done by two different *disciplines* (history of science, philosophy of science), which are quite jealous of

1. 'The Orthodox View of Theories', *Analyses of Theories and Methods of Physics and Psychology*, ed. Radner and Winokur, Minneapolis, 1970, p. 4.

their independence. But the question is not what distinctions a fertile mind can dream up when confronted with a complex process, or how some homogeneous material may be subdivided by accidents of history; the question is to what extent the distinction drawn reflects a real difference and whether science can advance without a strong interaction between the separated domains. (A river may be subdivided by national boundaries, but this does not make it a discontinuous entity.) Now there is, of course, a very noticeable difference between the rules of testing as 'reconstructed' by philosophers of science and the procedures which scientists use in actual research. This difference is apparent to the most superficial examination. On the other hand, a most superficial examination also shows that a determined application of the methods of criticism and proof which are said to belong to the context of justification, would wipe out science as we know it – and would never have permitted it to arise.[2] Conversely, the fact that science exists proves that these methods were frequently overruled. They were overruled by precisely those procedures which are now said to belong to the context of discovery. To express it differently: in the history of science, standards of justification often forbid moves that are caused by psychological, socio-economic-political and other 'external' conditions and science survives only because these moves are allowed to prevail. Thus the attempt 'to retrace the historical origins, the psychological genesis and development, the socio-political-economic conditions for the acceptance or rejection of scientific theories', far from being an enterprise that is entirely different from considerations of tests, actually leads to a criticism of these considerations – *provided* the two domains, historical research and discussion of test procedures, are not kept apart by fiat.

In a recent paper Feigl repeats his arguments and adds some further points. He is 'astonished that . . . scholars such as N. R. Hanson, Thomas Kuhn, Michael Polanyi, Paul Feyerabend, Sigmund Koch *et al.*, consider the distinction as invalid or at least misleading'.[3] And he points out that neither the psychology of invention nor any similarity, however great, between the sciences and the arts can show that it does not exist. In this he is certainly right. Even the most surprising stories about the manner in which scientists arrive at their theories cannot

2. Cf. the examples in Chapter 5.

3. 'Empiricism at Bay', MS, 1972, p. 2.

exclude the possibility that they proceed in an entirely different way once they have found them. *But this possibility is never realized.* Inventing theories and contemplating them in a relaxed and 'artistic' fashion, we often make moves that are forbidden by methodological rules. For example, we interpret the evidence so that it fits our fanciful ideas, we eliminate difficulties by *ad hoc* procedures, we push them aside, or we simply refuse to take them seriously. The activities which according to Feigl belong to the context of discovery are, therefore, not just *different* from what goes on in the context of justification, *they are in conflict with it.* The two contexts do not move along side by side, they frequently clash. And we are faced by the problem which context is to be given preferential treatment. This is part one of the argument. Now we have seen that in a case of conflict scientists occasionally choose the moves recommended by the context of justification, but they may also choose the moves that belong to the context of discovery and they often have excellent reasons for doing so. Indeed, science as we know it today could not exist without a frequent overruling of the context of justification. This is part two of the argument. The conclusion is clear. Part one shows that we do not just have a difference, we have an alternative. Part two shows that both sides of the alternative are equally important to science and that they must be given equal weight. Hence, we are not dealing with an alternative either, we are dealing with a single uniform domain of procedures all of which are equally important for the growth of science. This disposes of the distinction.

A similar argument applies to the ritual distinction between methodological *prescriptions* and historical *descriptions.* Methodology, it is said, deals with what *should* be done and cannot be criticized by reference to *what is.* But we must of course make sure that our prescriptions have a *point of attack* in the historical material, and we must also make sure that their determined application leads to desirable results. We make sure by considering (historical, sociological, physical, psychological, etc.) *tendencies and laws* which tell us what is possible and what is not possible under the given circumstances and thus separate feasible prescriptions from those which are going to lead into dead ends. Again, progress can be made only if the distinction between the *ought* and the *is* is regarded as a temporary device rather than as a fundamental boundary line.

A distinction which once may have had a point but which has now definitely lost it is the distinction between *observational* terms and *theoretical* terms. It is now generally admitted that this distinction is not as sharp as it was thought to be only a few decades ago. It is also admitted, in complete agreement with Neurath's original views, that *both* theories *and* observations can be abandoned: theories may be removed because of conflicting observations, observations may be removed for theoretical reasons. Finally, we have discovered that *learning* does not go from observation to theory but always involves both elements. Experience arises *together* with theoretical assumptions *not* before them, and an experience without theory is just as incomprehensible as is (allegedly) a theory without experience: eliminate part of the theoretical knowledge of a sensing subject and you have a person who is completely disoriented and incapable of carrying out the simplest action. Eliminate further knowledge and his sensory world (his 'observation language') will start disintegrating, colours and other simple sensations will disappear until he is in a stage even more primitive than a small child. A small child, on the other hand, does not possess a stable perceptual world which he uses for making sense of the theories put before him. Quite the contrary – he passes through various perceptual stages which are only loosely connected with each other (earlier stages *disappear* when new stages take over – see Chapter 17) and which embody all the theoretical knowledge available at the time. Moreover, the whole process starts only because the child reacts correctly towards signals, *interprets them correctly*, because he possesses means of interpretation even before he has experienced his first clear sensation.

All these discoveries cry out for a new terminology that no longer separates what is so intimately connected in the development both of the individual and of science at large. Yet the distinction between observation and theory is still upheld and is defended by almost all philosophers of science. But what is its point? Nobody will deny that the sentences of science can be classified into long sentences and short sentences, or that its statements can be classified into those which are intuitively obvious and others which are not. Nobody will deny that such distinctions *can be made*. But nobody will put great weight on them, or will even mention them, *for they do not now play any decisive role in the business of science*

(This was not always so. Intuitive plausibility, for example, was once thought to be a most important guide to the truth; it disappeared from methodology the very moment intuition was replaced by experience, and by formal considerations.) Does experience play such a role? It does not, as we have seen. Yet the inference that the distinction between theory and observation has now ceased to be relevant, is either not drawn or is explicitly rejected.[4] Let us take a step forward and let us abandon this last trace of dogmatism in science!

4. 'Neurath fails to give . . . rules [which distinguish empirical statements from others] and thus unwittingly throws empiricism overboard', K. R. Popper, *The Logic of Scientific Discovery*, New York and London, 1959, p. 97. For a more detailed argument concerning the observation–theory dichotomy cf. my essay 'Die Wissenschaftstheorie – eine bisher unbekannte Form des Irrsinns?', *Proceedings of the German Conference of Philosophy, Kiel, 1972*, Felix Meiner, Hamburg, 1973. 'Vagueness,' says Giedymin, 'seems to be the standard objection to the analytic–synthetic, observational–theoretical distinctions', *British Journal for the Philosophy of Science*, August 1970, p. 261. The objection is used by many authors, but it is certainly not used by me. See the considerations in the text above, 'Science without Experience', *Journal of Philosophy of Science*, 1969 (observational–theoretical) as well as the discussion remarks in Vol. I of the *Salzburg Studies in the Philosophy of Science*, Salzburg, 1967 (analytic–synthetic). My main objection is that the distinctions, while pleasing to simple minds, are *irrelevant* for the running of science and that the attempt to enforce them may arrest progress.

15

Finally, the discussion in Chapters 6–13 shows that Popper's version of Mill's pluralism is not in agreement with scientific practice and would destroy science as we know it. Given science, reason cannot be universal and unreason cannot be excluded. This feature of science calls for an anarchistic epistemology. The realization that science is not sacrosanct, and that the debate between science and myth has ceased without having been won by either side, further strengthens the case for anarchism.

Incommensurability, which I shall discuss next, is closely connected with the question of the rationality of science. Indeed one of the most general objections not merely to the *use* of incommensurable theories but even to the idea that *there are* such theories to be found in the history of science, is the fear that they would severely restrict the efficacy of traditional, non-dialectical *argument*. Let us, therefore, look a little more closely at the critical *standards* which, according to some, constitute the content of a 'rational' argument. More especially, let us look at the standards of the Popperian school with whose ratiomania we are here mainly concerned. This will prepare us for the final step in our discussion of the issue between law-and-order methodologies and anarchism in science.

Critical rationalism, which is the most liberal positivistic methodology in existence today, is either a meaningful idea or it is just a collection of slogans (such as 'truth', 'professional integrity', 'intellectual honesty', and so on) designed to intimidate yellow-bellied opponents (for who has the fortitude, or even the insight, to declare that 'truth' might be unimportant, and perhaps even undesirable?).

In the former case it must be possible to produce rules, standards, restrictions which permit us to separate critical behaviour (thinking, singing, writing of plays) from other types of behaviour so that we can *discover* irrational actions and *correct* them with the help of concrete suggestions. It is not difficult to produce the standards of rationality defended by the Popperian school.

These standards are standards of *criticism*: rational discussion consists in the attempt to criticize, and not in the attempt to prove or to make probable. Every step that protects a view from criticism, that makes it safe or 'well-founded', is a step away from rationality. Every step that makes it more vulnerable is welcome. In addition, it is recommended to abandon ideas which have been found wanting and it is forbidden to retain them in the face of strong and successful criticism unless one can present suitable counter-arguments. Develop your ideas so that they can be criticized; attack them relentlessly; do not try to protect them, but exhibit their weak spots; eliminate them as soon as such weak spots have become manifest – these are some of the rules put forth by our critical rationalists.

These rules become more definite and more detailed when we turn to the philosophy of science and, especially, to the philosophy of the natural sciences.

Within the natural sciences, criticism is connected with experiment and observation. The content of a theory consists in the sum total of those basic statements which contradict it; it is the class of its potential falsifiers. Increased content means increased vulnerability, hence theories of large content are to be preferred to theories of small content. Increase of content is welcome, decrease of content is to be avoided. A theory that contradicts an accepted basic statement must be given up. *Ad hoc* hypotheses are forbidden – and so on and so forth. A science, however, that accepts the rules of a critical empiricism of this kind will develop in the following manner.

We start with a *problem*, such as the problem of the planets at the time of Plato. This problem (which will be discussed in a somewhat idealized form) is not merely the result of *curiosity*, it is a *theoretical result*. It is due to the fact that certain *expectations* have been disappointed: on the one hand it seems to be clear that the stars must be divine, hence one expects them to behave in an orderly and lawful manner. On the other hand, one cannot find any easily discernible regularity. The planets, to all intents and purposes, move in a quite chaotic fashion. How can this fact be reconciled with the expectation and with the principles that underlie the expectation? Does it show that the expectation is mistaken? Or have we failed in our analysis of the facts? This is the problem.

It is important to see that the elements of the problem are not simply *given*. The 'fact' of irregularity, for example, is not accessible without further ado. It cannot be discovered by just anyone who has healthy eyes and a good mind. It is only through a certain expectation that it becomes an object of our attention. Or, to be more accurate, this fact of irregularity *exists* because there is an expectation of regularity. After all, the term 'irregularity' makes sense only if we have a rule. In our case the rule (which is a more specific part of the expectation) asserts circular motion with constant angular velocity. The fixed stars agree with this rule and so does the sun, if we trace its path relative to the fixed stars. The planets do not obey the rule, neither directly, with respect to the earth, nor indirectly, with respect to the fixed stars.

(In the problem we are examining now the rule is formulated explicitly and it can be discussed. This is not always the case. Recognizing a colour as red is made possible by deep-lying assumptions concerning the structure of our surroundings and recognition does not occur when these assumptions cease to be applicable.)

To sum up this part of the Popperian doctrine: research starts with a problem. The problem is the result of a conflict between an expectation and an observation which in turn is constituted by the expectation. It is clear that this doctrine differs from the doctrine of inductivism where objective facts enter a passive mind and leave their traces there. It was prepared by Kant, Poincaré, Dingler, and by Mill (*On Liberty*).

Having formulated a problem, one tries to *solve* it. Solving a problem means inventing a theory that is relevant, falsifiable (to a degree larger than any alternative), but not yet falsified. In the case mentioned above (planets at the time of Plato), the problem is: to find circular motions of constant angular velocity for the purpose of saving the planetary phenomena. The problem was solved by Eudoxos and Heracleides of Pontos.

Next comes the *criticism* of the theory that has been put forth in the attempt to solve the problem. Successful criticism removes the theory *once and for all* and creates a new problem, viz. to explain (a) why the theory was successful so far; (b) why it failed. Trying to solve *this* problem we need a new theory that reproduces the successful consequences of the older theory, denies its mistakes and makes additional predictions not made before. These are some of the *formal conditions* which *a suitable successor of a refuted theory* must satisfy. Adopting the conditions, one proceeds by conjecture and refutation from less general theories to more general theories and expands the content of human knowledge.

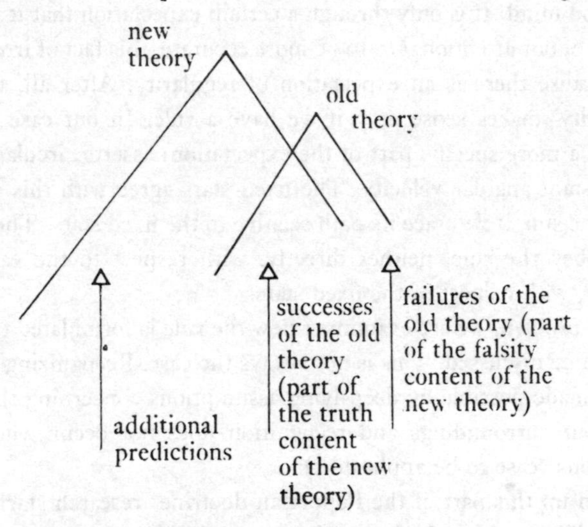

More and more facts are *discovered* (or constructed with the help of expectations) and are then explained by theories. There is no guarantee that man will solve every problem and replace every theory that has been refuted with a successor satisfying the formal conditions. The invention of theories depends on our talents and other fortuitous circumstances such as a satisfactory sex life. But as long as these talents hold out, the enclosed scheme is a correct account of the growth of a knowledge that satisfies the rules of critical rationalism.

Now at this point, one may raise two questions.

1. Is it *desirable* to live in accordance with the rules of a critical rationalism?

2. Is it *possible* to have both a science as we know it and these rules?

As far as I am concerned, the first question is far more important than the second. True, science and other increasingly depressing and narrow-minded institutions play an important part in our culture, and they occupy the centre of interest for many philosophers (most philosophers are opportunists). Thus the ideas of the Popperian school were obtained by generalizing solutions for methodological and epistemological problems. Critical rationalism arose from the attempt to solve Hume's problem and to understand the Einsteinian revolution, and it was then extended to politics and even to the conduct of one's private life. (Habermas and others therefore seem to be justified in calling Popper a positivist.) Such a procedure may satisfy a *school philosopher*, who looks at life through the spectacles of his own technical problems and recognizes hatred, love, happiness, only to the extent to which they occur in these problems. But if we consider the interests of *man* and, above all, the question of his freedom (freedom from hunger, despair, from the tyranny of constipated systems of thought and *not* the academic 'freedom of the will'), then we are proceeding in the worst possible fashion.

For is it not possible that science as we know it today, or a 'search for the truth' in the style of traditional philosophy, will create a monster? Is it not possible that it will harm man, turn him into a miserable, unfriendly, self-righteous mechanism without charm and humour? 'Is it not possible,' asks Kierkegaard, 'that my activity as an objective [or a critico-rational] observer of nature will weaken my strength as a human being?'[1] I suspect the answer to all these questions must be affirmative and I believe that a reform of the sciences that makes them more anarchistic and more subjective (in Kierkegaard's sense) is urgently needed.

But these are not the problems I want to discuss now. In the present essay I shall restrict myself to the second question and I shall ask: is it possible to have both a science as we know it and the rules of a critical rationalism as just described? And to *this* question the answer seems to be a firm and resounding *No*.

To start with we have seen, though rather briefly, that the actual development of institutions, ideas, practices, and so on, often *does not*

1. *Papirer*, ed. Heiberg, VII, Pt. I, sec. A, No. 182. Mill tries to show how scientific method can be understood as part of a theory of man and thus gives a positive answer to the question raised by Kierkegaard; cf. footnote 2 to Chapter 4.

start from a problem but rather from some irrelevant activity, such as playing, which, as a side effect, leads to developments which later on can be interpreted as solutions to unrealized problems.[2] Are such developments to be excluded? And, if we do exclude them, will this not considerably reduce the number of our adaptive reactions and the quality of our learning process?

Secondly, we have seen, in Chapters 8–12, that a *strict principle of falsification*, or a 'naive falsificationism' as Lakatos calls it,[3] would wipe out science as we know it and would never have permitted it to start.

The demand for *increased content* is not satisfied either. Theories which effect the overthrow of a comprehensive and well-entrenched point of view, and take over after its demise, are initially restricted to a fairly narrow domain of facts, to a series of paradigmatic phenomena which lend them support, and they are only slowly extended to other areas. This can be seen from historical examples (Chapters 8 and 9; footnote 1 of Chapter 9), and it is also plausible on general grounds: trying to develop a new theory, we must first take a *step back* from the evidence and reconsider the problem of observation (this was discussed in Chapter 12). Later on, of course, the theory is extended to other domains; but the mode of extension is only rarely determined by the elements that constitute the content of its predecessors. The slowly emerging conceptual apparatus of the theory *soon starts defining its own problems*, and earlier problems, facts, and observations are either forgotten or pushed aside as irrelevant (cf. the two examples in footnote 1 of Chapter 9 and the discussion towards the end of the next chapter). This is an entirely natural development, and quite unobjectionable. For why should an ideology be constrained by older problems which, at any rate, make sense only in the abandoned context and which look silly and unnatural now? Why should it even *consider* the 'facts' that gave rise to problems of this kind or played a role in their solution? Why should it not rather proceed in its own way, devising its own tasks and assembling its own domain of 'facts'? A comprehensive theory, after all,

2. Cf. the brief comments on the relation between idea and action in Chapter 1. For details cf. footnotes 31ff of 'Against Method', *Minnesota Studies*, Vol. 4, 1970.

3. 'Falsification and the Methodology of Scientific Research Programmes', *Criticism and the Growth of Knowledge*, ed. Lakatos and Musgrave, Cambridge, 1970, pp. 93 ff. ('Naive falsificationism' is here also called 'dogmatic'.)

is supposed to contain also an *ontology* that determines what exists and thus delimits the domain of possible facts and possible questions. The development of science agrees with these considerations. New views soon strike out in new directions and frown upon the older *problems* (what is the base upon which the earth rests? what is the specific weight of phlogiston? what is the absolute velocity of the earth?) and the older *facts* (most of the facts described in the *Malleus Maleficarum* – Chapter 9, footnote 1 – the facts of Voodoo – Chapter 4, footnote 8 – the properties of phlogiston or those of the ether) which so much exercised the minds of earlier thinkers. And where they *do* pay attention to preceding theories, they try to accommodate their factual core in the manner already described, with the help of *ad hoc* hypotheses, *ad hoc* approximations, redefinition of terms, or by simply *asserting*, without any more detailed study of the matter, that the core 'follows from' the new basic principles.[4] They are 'grafted on to older programmes with which they [are] blatantly inconsistent'.[5]

The result of all these procedures is an interesting *epistemological illusion*: the *imagined* content of the earlier theories (which is the intersection of the remembered consequences of these theories with the newly recognized domain of problems and facts) *shrinks* and may decrease to such an extent that it becomes smaller than the *imagined* content of the new ideologies (which are the actual consequences of these ideologies *plus* all those 'facts', laws, principles which are tied to them by *ad hoc* hypotheses, *ad hoc* approximations or by the say-so of some influential physicist or philosopher of science – and which properly belong to the predecessor). Comparing the old and the new it thus *appears* that the relation of empirical contents is like this

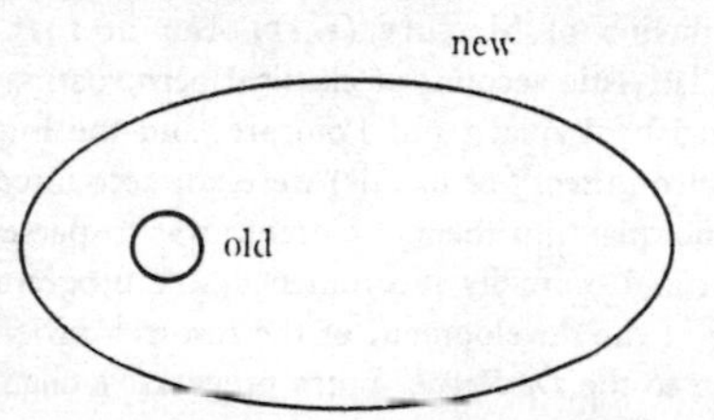

4. 'Einstein's theory is better than ... Newton's theory *anno 1916* ... *because* it explained everything that Newton's theory had successfully explained ...', Lakatos, op. cit., p. 124.

5. Lakatos, discussing Copernicus and Bohr, ibid., p. 143.

or, perhaps, like this

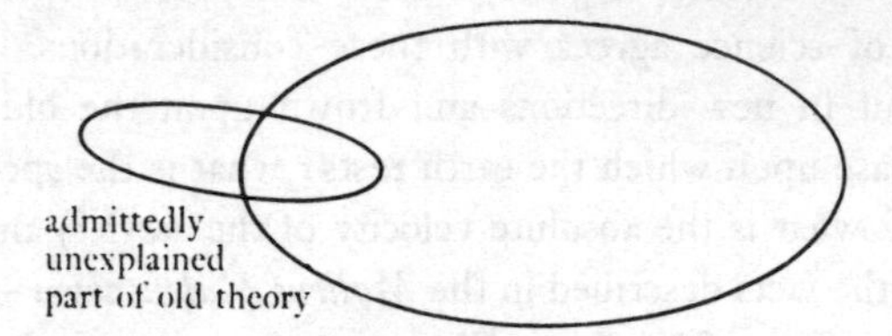

while in actual fact it is much more like this

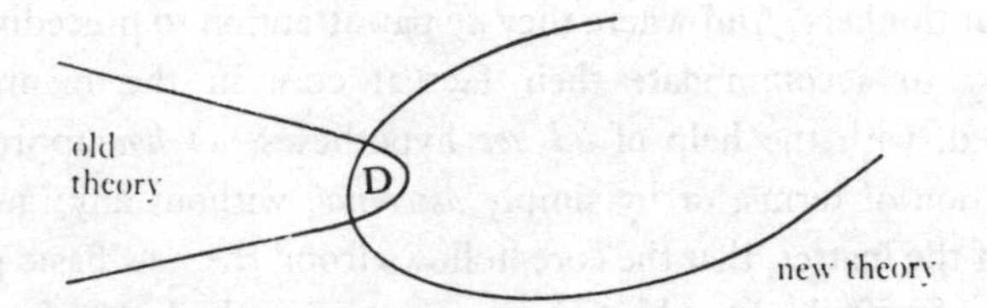

domain D representing the problems and facts of the old theory which are still remembered and which have been distorted so as to fit into the new framework. It is this illusion which is responsible for the persistent survival of the demand for increased content.[6]

Finally, we have by now seen quite distinctly the need for *ad hoc hypotheses*: *ad hoc* hypotheses and *ad hoc* approximations create a tentative area of contact between 'facts' and those parts of a new view which seem capable of explaining them, at some time in the future and after addition of much further material. They specify possible explananda and explanatia, and thus determine the direction of future research.

6. This illusion occurs even in Elie Zahar's excellent paper on the development from Lorentz to Einstein. According to Zahar, Einstein superseded Lorentz with the explanation of the perihelion of Mercury (1915). But in 1915 nobody had as yet succeeded in giving a relativistic account of classical perturbation theory to the degree of approximation reached by Laplace and Poincaré, and the implications of Lorentz on the atomic level (electron theory of metals) were not accounted for either, but were gradually replaced by the quantum theory: Lorentz was 'superseded' not by one, but by at least two different and mutually incommensurable programmes. Lakatos, in his excellent reconstruction of the development of the research programme of Copernicus from the *Commentariolus* to the *De Revol.*, notes progressive changes but only because he omits the dynamical and the optical problems and concentrates on kinematics, pure and simple. Small wonder that both Zahar and Lakatos are under the impression that the content condition is still satisfied. Cf. also my short note 'Zahar on Einstein', in the *British Journal for the Philosophy of Science*, March, 1974.

They may have to be retained forever if the new framework is partly unfinished (this happened in the case of the quantum theory, which needs the classical concepts to turn it into a complete theory).[7] Or they are incorporated into the new theory as theorems, leading to a redefinition of the basic terms of the preceding ideology (this happened in the cases of Galileo and of the theory of relativity). The demand that the truth-content of the earlier theory *as conceived while the earlier theory reigned supreme* be included in the truth-content of the successor is violated in either case.

To sum up: wherever we look, whatever examples we consider, we see that the principles of critical rationalism (take falsifications seriously; increase content; avoid *ad hoc* hypotheses; 'be honest' – whatever *that* means; and so on) and, *a fortiori*, the principles of logical empiricism (be precise; base your theories on measurements; avoid vague and unstable ideas; and so on) give an inadequate account of the past development of science and are liable to hinder science in the future. They give an inadequate account of science because science is much more 'sloppy' and 'irrational' than its methodological image. And, they are liable to hinder it, because the attempt to make science more 'rational' and more precise is bound to wipe it out, as we have seen. The difference between science and methodology which is such an obvious fact of history, therefore, indicates a weakness of the latter, and perhaps of the 'laws of reason' as well. For what appears as 'sloppiness', 'chaos' or 'opportunism' when compared with such laws has a most important function in the development of those very theories which we today regard as essential parts of our knowledge of nature. *These 'deviations', these 'errors', are preconditions of progress.* They permit knowledge to survive in the complex and difficult world which we inhabit, they permit *us* to remain free and happy agents. Without 'chaos', no knowledge. Without a frequent dismissal of reason, no progress. Ideas which today form the very basis of science exist only because there were such things as prejudice, conceit, passion; because these things *opposed reason*; and because they *were permitted to have their way*. We have to conclude then, that *even within* science reason cannot and should not be allowed to be comprehensive and that it must often be overruled, or eliminated, in favour of other

7. Cf. 'Problems of Empiricism', Part II, op. cit., sections 9 and 10.

agencies. There is not a single rule that remains valid under all circumstances and not a single agency to which appeal can always be made.

Now we must remember that this conclusion has been drawn *on condition* that science as we know it today remains unchanged and that the procedures it uses be permitted to determine its future development as well. *Given* science, reason cannot be universal and unreason cannot be excluded. This peculiar feature of the development of science strongly supports an anarchistic epistemology. But science is not sacrosanct. The restrictions it imposes (and there are many such restrictions though it is not easy to spell them out) are not necessary in order to have general coherent and successful views about the world. There are myths, there are the dogmas of theology, there is metaphysics, and there are many other ways of constructing a world-view. It is clear that a fruitful exchange between science and such 'non-scientific' world-views will be in even greater need of anarchism than is science itself. Thus anarchism is not only *possible*, it is *necessary* both for the internal progress of science and for the development of our culture as a whole. And, Reason, at last, joins all those other abstract monsters such as Obligation, Duty, Morality, Truth and their more concrete predecessors, the Gods, which were once used to intimidate man and restrict his free and happy development: it withers away. . . .

16

Weil er uns sonst niederhaut,
Preisen wir ihn alle laut.

From the introductory chorus
of Nestroy's *Judith und Holofernes.*

Even the ingenious attempt of Lakatos to construct a methodology that (a) does not issue orders and yet (b) puts restrictions upon our knowledge-increasing activities does not escape this conclusion. For Lakatos' philosophy appears liberal only because it is an anarchism in disguise. *And his standards which are abstracted from modern science cannot be regarded as neutral arbiters in the issue between modern science and Aristotelian science, myth, magic, religion, etc.*

This would have been the end of my essay in defence of an epistemological anarchism had it not been for the fact that the drive for law and order in science and philosophy proceeds undiminished and that it has found a new and most effective champion in the person of Imre Lakatos. The task which Lakatos sets himself – to increase the number of the Friends of Reason and to reassure doubtful and apprehensive rationalists – is in some way not at all difficult. For it needs only a few well-placed phrases to put the fear of Chaos into the most enlightened audience, and to make them yearn for simple rules and simple dogmas which they can follow without having to reconsider matters at every turn. Some of the most outspoken anarchists rely on science and reason, even on induction, as we have seen.[1] And the younger generation who are so vociferous in their contempt for authority are not prepared to live without the authority of Reason either. I must confess that this almost universal urge for 'objective' guidance is somewhat of a puzzle to me. I am not surprised

1. Cf. *Introduction*, text to footnote 12.

when experts who are advanced in years, who have a reputation to uphold (or to get quickly, before they die), and who quite naturally confound knowledge with mental rigor mortis look askance at attempts to loosen up science, or to demonstrate that *great* science (which is not the science of the schools, nor the science of the Rand Corporation, and certainly not the science of Fallowfield or of the London School of Economics) is an intellectual adventure that knows of no limits, and recognizes no rules, not even the rules of logic. But I do find it a little astonishing to see with what fervour students and other non-initiates cling to stale phrases and decrepit principles as if a situation in which they bear the responsibility for *every* action and are the original cause for *every* regularity of the mind were quite unbearable to them. Considering this attitude an appeal to reason is bound to find an attentive audience even if it is entirely devoid of reason itself. This is what I meant when saying that the task which Lakatos sets himself is in some ways not at all difficult. But the task is very difficult in another respect: it is very difficult to overcome the obstacles to reason which recent research has discovered, and to develop a form of rationalism that can cope with them. Yet this is exactly what Lakatos tries to do. Let us see how he proceeds!

Lakatos criticizes the existent methodologies and he arrives at a result that is almost identical with mine. Considering the manner in which theories are eliminated he writes: 'If we look at the history of science, if we try to see how some of the most celebrated falsifications happened, we have to come to the conclusion that either some of them are plainly irrational, or that they rest on rationality principles different from those which we have just discussed.'[2] The 'rationality principles we have just discussed' are the principles of critical rationalism as outlined in the preceding section; but Lakatos would be prepared to extend his observation to other methodologies, and to events other than falsification.[3] He is one of the very few thinkers who have noticed the tremendous gulf that exists between various *images* of science and the 'real thing'; and he has also realized that the attempt to *reform* the sciences by bring-

2. Lakatos, 'Falsification and the Methodology of Research Programmes' in *Criticism and the Growth of Knowledge*, Cambridge, 1970, p. 114. Hereafter 'Falsification'.

3. Cf. 'Falsification', p. 104, for consistency.

ing them closer to the image is bound to damage them and may even destroy them. With this result I certainly agree.

I also agree with two suggestions which form an essential part of Lakatos' theory of science. The first suggestion is that methodology must grant 'a breathing space'[4] to the ideas we wish to consider. Given a new theory, we must not at once use the customary standards for deciding about its survival. Neither blatant internal inconsistencies, nor obvious lack of empirical content, nor massive conflict with experimental results should prevent us from retaining and elaborating a point of view that pleases us for some reason or other.[5] It is the *evolution* of a theory over long periods of time and not its shape at a particular moment that counts in our methodological appraisals. This suggestion removes most of the objections I have raised in preceding chapters.

Secondly, Lakatos suggests that methodological standards are not beyond criticism. They can be examined, improved, replaced by better standards. The examination is not abstract, but makes use of *historical data*: historical data play a decisive role in the debate between rival methodologies. This second suggestion separates Lakatos and myself from the logicians who regard an appeal to history as 'a method of very poor efficiency',[6] and who believe that methodology should be run on the basis of simple models only. (Many logicians do not even see the problem; they take it for granted that constructing formal systems and playing with them is the only legitimate way of understanding scientific change.)[7]

4. 'History of Science and its Rational Reconstructions' in *Boston Studies for the Philosophy of Science*, Vol. VIII, p. 113. Hereafter 'History'.

5. Examples are: *lack of content* – the atomic theory throughout the ages; the idea of the motion of the earth of Philolaus; *inconsistency* – Bohr's programme (cf. 'Falsification', pp. 138ff); *massive conflict with experimental results* – the idea of the motion of the earth as described in Chapters 6ff above; Prout's theory as described in 'Falsification', pp. 138ff.

6. R. Carnap, *Logical Foundations of Probability*, Chicago, 1950, p. 217.

7. R. Carnap, p. 202, draws a distinction between logical and methodological problems and he warns us that the problems of psychology and sociology which accompany the application of systems of inductive logic 'should not be regarded as difficulties of inductive logic itself' (p. 254). He thus seems to realize the need for a factual evaluation of *applied* inductive logic. But this factual evaluation is carried out in the same abstract way that led to the construction of an inductive logic in the first place. In addition to a 'simple universe' without which the business of inductive logic could not even get started we also use 'an observer X with a simplified biography' (p. 213). Now I do not

My quarrel with Lakatos concerns the standards he recommends, his evaluation of modern science (in comparison with, say, myth or Aristotelian science), his contention that he has proceeded 'rationally', as well as the particular historical data he uses in his discussion of methodologies. I start with an account of the first item on the list.

When a new theory or a new idea enters the scene it is usually somewhat inarticulate, it contains contradictions, the relation to facts is unclear, ambiguities abound. The theory is full of faults. However, it can be developed, and it may improve. The natural unit of methodological appraisals is therefore not a single theory, but a succession of theories, or a *research programme*; and we do not judge the *state* in which a research programme finds itself at a particular moment, we judge its *history*, preferably in comparison with the history of rival programmes.

According to Lakatos the judgements are of the following kind: 'A research programme is said to be *progressing* as long as its theoretical growth anticipates its empirical growth, that is as long as it keeps predicting novel facts with some success . . .; it is *stagnating* if its theoretical growth lags behind its empirical growth, that is, as long as it gives only *post hoc* explanations of either chance discoveries or of facts anticipated by, and discovered in a rival programme'.[8] A stagnating programme may *degenerate* further until it contains nothing but 'solemn reassertions'

---

object to the procedure of abstraction itself. But when abstracting from a particular feature of science we should make sure that science can exist without it, that an activity, not necessarily science, that lacks it, is (physically, historically, psychologically) *possible*; and we should also take care to *restore* the omitted feature when the abstract debate has come to an end. (In this respect scientists and philosophers of science act in very different ways. The physicist who has used geometry [which disregards weight] to calculate some properties of a physical object puts the weight back after he has finished his calculations. Not once does he assume that the world is full of weightless shapes. The philosopher who has used deductive logic [which disregards contradictions] to ascertain some properties of a scientific argument never puts the contradictions back into the argument after he has finished *his* work and he assumes that the world is full of self-consistent theoretical systems.) Now the only way of finding out whether a certain feature is necessary for science is to carry out a *functional study* of this feature (in the sense of modern anthropology) that examines its role in the growth of science. This leads us right back into history where we find the data for such a study. Without them there is no way of knowing whether 'the roundabout way through an abstract scheme' is indeed 'the best way' of doing methodology (p. 217) and no possibility of judging the scheme that has actually been proposed.

8. 'History', p. 100.

of the original position coupled with a repetition, in its own terms, of (the successes of) rival programmes.[9] Judgements of this kind are central to the methodology Lakatos wishes to defend. They *describe* the situation in which a scientist finds himself. *They do not yet advise him how to proceed.*

Considering a research programme in an advanced state of degeneration one will feel the urge to abandon it, and to replace it by a more progressive rival. This is an entirely legitimate move. *But it is also legitimate to do the opposite* and to retain the programme. For any attempt to demand its removal on the basis of a *rule* can be criticized by arguments almost identical with the arguments that led to the 'securing of a breathing space' in the first place: if it is unwise to reject faulty theories the moment they are born because they might grow and improve, then it is also unwise to reject research programmes on a downward trend because they might recover and might attain unforeseen splendour (the butterfly emerges when the caterpillar has reached its lowest stage of degeneration). Hence, one cannot *rationally* criticize a scientist who sticks to a degenerating programme and there is no *rational* way of showing that his actions are unreasonable.

Lakatos agrees with this. He emphasizes that one 'may rationally stick to a degenerating programme until it is overtaken by a rival *and even after*'[10] 'programmes may get out of degenerating troughs'.[11] It is true that his rhetoric frequently carries him much further, showing that he has not yet become accustomed to his own liberal proposals.[12] But

9. ibid., p. 105; details in 'Falsification', pp. 116ff.

10. ibid., p. 104.

11. 'Falsification', p. 164.

12. 'I give rules for the "elimination" of whole research programmes', 'History', p. 100 – note the ambiguity introduced by the quotation marks. Occasionally the restrictions are introduced in a different way, by denying the 'rationality' of certain procedures. 'It is perfectly rational to play a risky game', says Lakatos ('History', p. 104), 'what is irrational is to deceive oneself about the risk': one can do whatever one wants to do if one occasionally remembers (or recites?) the standards *which, incidentally, say nothing about risks, or the size of risks.* Speaking about risks either involves a *cosmological* assumption (Nature only rarely permits research programmes to behave like caterpillars) or a *sociological* assumption (*institutions* only rarely permit degenerating programmes to survive). Lakatos in passing ('History', p. 101) admits the need for such additional assumptions: only they 'can turn science from a mere game into an epistemologically rational exercise'. But he does not *discuss* them in detail and those he takes for granted are very doubtful, to say the least. Take the cosmological assumption

when the issue arises in explicit form, then the answer is clear: the methodology of research programmes provides *standards* that aid the scientist in evaluating the historical situation in which he makes his decisions; it does not contain *rules* that tell him what to do.[13]

The methodology of research programmes thus differs radically from inductivism, falsificationism and from other and even more paternalistic philosophies. Inductivism demands that theories that lack empirical support *be removed*. Falsificationism demands that theories that lack excess empirical content over their predecessors *be removed*. Everyone demands that inconsistent theories, or theories with low empirical content, *be removed*. The methodology of research programmes neither *does* contain such demands nor *can* it contain them, as we have seen. Its rationale – 'to provide a breathing space' – and the arguments that established the need for more liberal standards make it impossible to specify conditions in which a research programme *must* be abandoned, or when it becomes *irrational* to continue supporting it. *Any* choice of the scientist is rational, because it is compatible with the standards. 'Reason' no longer influences the actions of the scientist. (But it provides terminology for describing the results of these actions.)

Let me repeat the steps which lead to this surprising result. Step one is the definition of reason (the 'rationality theory') that is accepted by Lakatos. It is contained in his standards for the comparative appraisal of research programmes. Step two is the observation[14] that the standards, taken by themselves, have no heuristic force. Reason as defined by Lakatos does not *directly* guide the actions of the scientist. Given this reason and nothing else, 'anything goes'. It follows that there is no

---

which I have just mentioned. It is quite interesting, and it certainly deserves being studied in greater detail. Such a study, I venture to suggest, would reveal that the research programme corresponding to it is now in a degenerate phase. (To see this one, needs only to consider *anomalies* such as the Copernican Revolution, the revival of the atomic theory, the revival of the assumption of celestial influences as well as the *ad hoc* adaptations of these anomalies that are reflected in the 'epistemological illusion' described in Chapter 15.) The sociological assumption, on the other hand, is certainly true – which means that given a world in which the cosmological assumption is false we shall forever be prevented from finding the truth.

13. 'History', p. 104, last four lines.

14. Which is repeatedly emphasized by Lakatos himself: 'History', pp. 92, 104, footnotes 2, 57, *et al.*

'rationally' describable difference between Lakatos and myself, always taking Lakatos' standards as a measure of reason. However, there is certainly a great difference in *rhetorics*: and we also differ in our attitude towards the 'freedom' of research[15] that emerges from our 'standards'. I now take a closer look at these differences.

The hallmark of *political anarchism* is its opposition to the established order of things: to the state, its institutions, the ideologies that support and glorify these institutions. The established order must be destroyed so that human spontaneity may come to the fore and exercise its right of freely initiating action, of freely choosing what it thinks is best. Occasionally one wishes to overcome not just some social circumstances but the entire physical world which is seen as being corrupt, unreal, transient, and of no importance. This *religious* or *escatological* anarchism denies not only social laws, but moral, physical and perceptual laws as well and it envisages a mode of existence that is no longer tied to the body, its reactions, and its needs. *Violence*, whether political or spiritual, plays an important role in almost all forms of anarchism. Violence is *necessary* to overcome the impediments erected by a well-organised society, or by one's own modes of behaviour (perception, thought, etc.), and it is *beneficial* for the individual, for it releases one's energies and makes one realize the powers at one's disposal. Free associations where everyone does what best suits their talents replace the petrified institutions of the day, no function must be allowed to become fixed – 'the commander of yesterday can become a subordinate of tomorrow'.[16] Teaching is to be based on curiosity and not on command, the 'teacher' is called upon to further this curiosity and not to rely on any fixed method. Spontaneity reigns supreme, in thought (perception) as well as in action.

One of the remarkable characteristics of post-enlightenment political anarchism is its faith in the 'natural reason' of the human race and its respect for science. This respect is only rarely an opportunistic move – one

15. One should remember that the debate is about methodological rules only and that 'freedom' now means freedom *vis-à-vis* such rules. The scientist is still restricted by the properties of his instruments, the amount of money available, the intelligence of his assistants, the attitude of his colleagues, his playmates – he or she is restricted by innumerable physical, physiological, sociological, historical constraints. The methodology of research programmes (and the epistemological anarchism I advocate) removes only the methodological constraints.

16. Bakunin, *Oeuvres*, Vol. II, p. 297.

recognizes an ally and compliments him to keep him happy. Most of the time it is based on the genuine conviction that pure unadulterated science gives a true account of man and the world and produces powerful ideological weapons in the fight against the sham orders of the day.

Today this naive and almost childlike trust in science is endangered by two developments.

The first development is the rise of new kinds of scientific institutions. As opposed to its immediate predecessor, late 20th-century science has given up all philosophical pretensions and has become a powerful *business* that shapes the mentality of its practitioners. Good payment, good standing with the boss and the colleagues in their 'unit' are the chief aims of these human ants who excel in the solution of tiny problems but who cannot make sense of anything transcending their domain of competence. Humanitarian considerations are at a minimum[17] and so is any form of progressiveness that goes beyond local improvements. The most glorious achievements of the past are used not as instruments of enlightenment but as means of intimidation as is seen from some recent debates concerning the theory of evolution. Let somebody make a great step forward – and the profession is bound to turn it into a club for beating people into submission.

The second development concerns the alleged authority of the *products* of this ever-changing enterprise. Scientific laws were once thought to be well established and irrevocable. The scientist discovers facts and laws and constantly increases the amount of *safe* and *indubitable* knowledge. Today we have recognized, mainly as a result of the work of Mill, Mach, Boltzmann, Duhem and others, that science cannot give any such guarantees. Scientific laws can be revised, they often turn out to be not just locally incorrect but entirely false, making assertions about entities that never existed. There are revolutions that leave no stone unturned, no principle unchallenged. Unpleasant in appearance, untrust-

17. 'The desire to alleviate suffering is of small value in research', writes a modern Frankenstein, Dr Szentgyorgi, in *Lancet i*, 1961, p. 1394 (from a talk given at an international medical congress). 'Such a person should be advised to work for charity. Research wants egoists, damned egoists, who seek their own pleasure and satisfaction, but find it in solving the puzzles of nature.' For the effects of this attitude on the activities of physicians cf. M. H. Pappworth, *Human Guinea Pigs*, Boston, 1965. For some effects in psychiatry cf. D. L. Rosenhan, *Science 179*, 1973, pp. 250ff.

worthy in its results, science has ceased to be an ally of the anarchist and has become a problem. Should he abandon it? Should he use it? What should he do with it? That is the question. Epistemological anarchism gives an answer to this question. It is in line with the remaining tenets of anarchism and it removes the last hardened elements.

Epistemological anarchism differs both from scepticism and from political (religious) anarchism. While the sceptic either regards every view as equally good, or as equally bad, or desists from making such judgements altogether, the epistemological anarchist has no compunction to defend the most trite, or the most outrageous statement. While the political or the religious anarchist wants to remove a certain form of life, the epistemological anarchist may want to defend it, for he has no everlasting loyalty to, and no everlasting aversion against, any institution or any ideology. Like the Dadaist, whom he resembles much more than he resembles the political anarchist, he 'not only has no programme, [he is] against all programmes',[18] though he will on occasions be the most vociferous defender of the *status quo*, or of his opponents: 'to be a true Dadaist, one must also be an anti-Dadaist'. His aims remain stable, or change as a result of argument, or of boredom, or of a conversion experience, or to impress a mistress, and so on. Given some aim, he may try to approach it with the help of organized groups, or alone; he may use reason, emotion, ridicule, an 'attitude of serious concern' and whatever other means have been invented by humans to get the better of their fellow men. His favourite pastime is to confuse rationalists by inventing compelling reasons for unreasonable doctrines. There is no view, however 'absurd' or 'immoral', he refuses to consider or to act upon, and no method is regarded as indispensable. The one thing he opposes positively and absolutely are universal standards, universal laws, universal ideas such as 'Truth', 'Reason', 'Justice', 'Love' and the behaviour they bring along, though he does not deny that it is often good policy to act as if such laws (such standards, such ideas) existed, and as if he believed in them. He may approach the religious anarchist in his opposition to science and the material world, he may outdo any Nobel Prize winner in his vigorous defence of scientific purity. He has no objection to

18. For this and the two following quotations cf. Hans Richter, *Dada – Art an Anti-Art*, London, 1965.

regarding the fabric of the world as described by science and revealed by his senses as a chimera that either conceals a deeper and, perhaps, spiritual reality, or as a mere web of dreams that reveals, and conceals, nothing. He takes great interest in procedures, phenomena and experiences such as those reported by Carlos Castaneda,[19] which indicate that perceptions can be arranged in highly unusual ways and that the choice of a particular arrangement as 'corresponding to reality', while not arbitrary (it almost always depends on traditions), is certainly not more 'rational' or more 'objective' than the choice of another arrangement: Rabbi Akiba, who in ecstatic trance rises from one celestial sphere to the next and still higher and who finally comes face to face with God in all his Splendour,[20] *makes genuine observations* once we decide to accept his way of life as a measure of reality, and his mind is as independent of his body as the chosen observations tell him.[21] Applying this point of view to a specific subject such as science, the epistemological anarchist finds that its accepted development (e.g. from the Closed World to the 'Infinite Universe') occurred only because the practitioners unwittingly used his philosophy within the confines of their trade – they succeeded because they did not permit themselves to be bound by 'laws of reason', 'stand-

19. *The Teachings of Don Juan*, New York, 1968. Like other 'experiments' these experiences are prepared in two ways. There is a long-term preparation and a short-term preparation. The long-term preparation consists of a series of personality tests, explanations of the purpose of the tests as well as of their results, drug-induced hallucinogenic states and it is summarized in a complex and most interesting theory of knowledge, or paths to knowledge (op. cit., pp. 79ff). The short-term preparation consists in inducing the hallucinogenic state and in particular instructions given (cf. the instructions for becoming a raven, op. cit., pp. 172ff). Long-term and short-term preparations taken together give meaning to the experiences and unite them into a single and coherent world that is more or less strongly connected with the everyday world but that is occasionally completely separate. Criteria may differ in both cases, of course, but there is no objective way of deciding between them unless one finds a 'superworld' that includes experiences of both kinds. And even in this case we need criteria for evaluating the experiences and we have to decide between various possibilities.

20. Cf. W. Bousset, 'Die Himmelsreise der Seele', *Archiv für Religionswissenschaft*, Bd. 4, 1901, pp. 136ff. Reprint Darmstadt, 1961, p. 14.

21. 'Command your soul to be in India, to cross the ocean; in a moment it will be done. And if you wish to break through the vault of the universe and to contemplate what is beyond – if there is anything beyond the world – you may do it', *Corpus Hermeticum*, XII, quoted after Festugière, *La Révélation d'Hermès Trismégiste*, Paris, 1950, Vol. I, p. 147.

ards of rationality', or 'immutable laws of nature'. Underneath all this outrage lies his conviction that man will cease to be a slave and gain a dignity that is more than an exercise in cautious conformism only when he becomes capable of stepping outside the most fundamental categories and convictions, including those which allegedly make him human. 'The realisation that reason and anti-reason, sense and nonsense, design and chance, consciousness and unconsciousness [and, I would add, humanitarianism and anti-humanitarianism] belong together as a necessary part of a whole – this was the central message of Dada,' writes Hans Richter. The epistemological anarchist agrees, though he would not express himself in such a constipated manner. There is no room, in the present essay, to pursue all the implications of this radical view which is reasonable in the sense that every move it recommends can be defended with the help of the most beautiful arguments (after all, reason *is* the slave of passions). Instead, I shall try to show how an epistemological anarchist might act in specific problem situations, assuming he has temporarily decided to choose a certain aim, and to accept a certain description of the 'state of the world'.

Let us imagine that he lives at the beginning of the 17th century and has just become acquainted with Copernicus' main work. What will be his attitude? What moves will he recommend? What moves will he oppose? What will he say? What he will say depends on his interests, on the 'social laws', the social philosophy, the opinions concerning the contemporary scene he has decided to adopt *for the time being*. There are innumerable ways in which he may justify these laws, these opinions, this philosophy to those who demand a justification, or at least an argument. We are not interested in such justification and such arguments.

Assume furthermore that our anarchist is interested not just in technical developments but in *social peace*, and he realizes that social peace can be disturbed by developments in recondite fields (note that the words 'interest' and 'realize' and all further descriptions of his activity are common-sense descriptions which entail a methodological attitude not shared by the anarchist: *he* is like an undercover agent who works on both sides of the fence). He will then study the ideological potential of Copernicanism given the existence of new and somewhat restless classes who might claim Copernicus as a supporter of their interests but who can also be reached, and *tamed*, by arguments. Being convinced of the 'rationality' of his opponents (provided the reasons are not given in dry and scholarly language) he will prepare entertaining tracts ('entertaining' from the point of view of his readers), emphasizing the weak points of the Copernican theory, and he will organize the livelier intellectuals for the most efficient completion of this task. He may well be successful – for 'it is very difficult to defeat a research programme supported by talented, imaginative scientists':[22] 'if two teams, pursuing rival research programmes compete, the one with more creative talent [and, so one should add, with more insight into social conditions and the psyche of the opponents] is more likely to succeed . . . the direction of science is determined primarily by human creative imagination and not by the universe of facts, that surrounds us.'[23] He may proceed more directly and defend the ideal of *stability* that underlies the Aristotelian point of view and that still appeals to sizeable groups of the total population. This is how, playing the game of some rationalists and using social laws as temporary levers, the anarchist may rationally defeat the urge for progress of other rationalists.

It is interesting to see that Cardinal Bellarmine, though by no means an anarchist, was guided by considerations very similar to those just outlined: he wants social peace. 'Galileo did not himself show much concern for the common, ignorant people, the "herd" as he called them, in his rather snobbish attitude to all who were not great mathematicians and experimentalists of his own type. Even if, as he suggested, they should lose their faith through being told that the earth was speeding round the sun at a rate of eighteen miles per second, still Copernicanism must be preached in season and out of season. The common man . . . was a person very dear to the heart of Bellarmine, and he could not understand Galileo's headlong precipitancy in forcing an issue that might trouble the faith of the simple when he could so easily have kept his intuitions, as scientists do today, for debate and quiet study among his peers. Bellarmine was surely entitled to ask for some more solid proof than the moons of Jupiter, the phases of Venus, and the spots on the Sun, all of which fitted perfectly well into Tycho Brahe's system, while leaving the earth stationary. . . . [T]his was the system adopted

22. 'Falsification', p. 158.   23. ibid. p. 187.

by the Jesuit astronomers. . . .'[24] (Unfortunately [or fortunately?] these astronomers rested content with raising difficulties and adapting discoveries that had been made by somebody else, they did not realize the propagandistic value of predictions and dramatic shows, nor did they utilize the intellectual and the social powers of the newly rising classes. *They lost by default.*)

Assume on the other hand our anarchist detests the emotional, the intellectual, the social bonds to which his contemporaries are subjected, that he regards them as a hindrance rather than as the presupposition of a happy and fulfilling life and that, being an intellectual and not a general, or a bishop, he prefers to change the situation while remaining seated in his study. In this case he will look for views which are opposed to some fundamental assumptions of the orthodox ideology and which can be used as intellectual *levers* for overthrowing this ideology. He will realize that abstract ideas can become such levers only if they are part of a practice, a 'form of life' that (a) *connects them* with influential events and (b) has itself some social *influence* – otherwise they are disregarded or laughed out of court as signs of intellectual sophistry and remoteness. There must be a tradition that can absorb the new ideas, use them, elaborate them, and this tradition must be respected among influential people, powerful classes, etc. Our anarchist may decide that the Copernican point of view is a potential lever of the kind he wants and he may look around for means of making it more efficient. The first subject, or 'form of life', he encounters in his search is, of course, astronomy and, within astronomy, the demand for better tables, better values of constants, better means of fixing the calendar. Progress in this direction would strengthen the Copernican point of view and thus strengthen his lever. But even the greatest predictive success is defused at once by a familiar theory that is also part of astronomy, and that seems to have the support of the Great Copernicus himself[25]: astronomical theories are *instruments* of prediction; their success does not tell us anything about the actual structure of the universe; problems of this kind are settled by *physics*, on the basis of simple observations. This 'instrumentalist

24. James Broderick, S.J., *Robert Bellarmine, Saint and Scholar*, London, 1961, pp. 366ff.

25. 'Many otherwise perspicacious readers of the *Revolutions* were fooled by Osiander's mutilation', E. Rosen, *Three Copernican Treatises*, New York, 1971, p. 40.

view' is not only an important part of the tradition he wants to use, it can also be supported by observations different from those that support physics: look at Venus, or at Mars, and you will see that they increase and decrease in size in a way very different from the increase and decrease demanded by the Copernican arrangement of their paths.[26] This shows that additional means are needed for strengthening the view that is to explode the *status quo*. Means that cannot so easily be interpreted in an instrumentalistic manner. So our anarchist changes his method. He disregards the complications of planetary astronomy,[27] he lets the planets move in simple circles and he tries to find more direct signs of the truth of the Copernican view. By a stroke of luck he has heard of the telescope. It seems to be an important aid in the art of war, it has caught the attention of the public, it is surrounded by mystery, one is prepared to trust it or, better, those artisans who from a close acquaintance with lenses have some *practical* experience with contraptions of this kind are prepared to trust it. Public exhibitions are arranged. Things are seen that cannot be seen with the naked eye and whose nature is known independently – towers, walls, ships, etc. Nobody doubts that the instrument shows how things really are. The stage is set. And now the telescope is directed towards the sky. Numerous puzzling phenomena appear, some of them absurd, some contradictory, *some giving direct support to the Copernican view*. Even the most sophisticated optical argument cannot stop the rising conviction that a new age of knowledge has started and that the old stories about the sky are just – stories. The conviction is especially strong among those who have advanced knowledge in a practical way, without involved terminology, and who are convinced that university physics is a collection of words rather than a knowledge of things (remember the Puritans' contempt for useless speculation). Asked for a theoretical justification our anarchist, remembering the law of uneven development, will use rags of argument in a shamelessly propagandistic fashion. Very often the enthusiasm for the new views is strong enough to make additional propaganda unnecessary: 'It was fortunate for these men that their sympathies sometimes obscured their critical vision', writes Albert Schweitzer about analogous developments in

26. Cf. Appendix 1, p. 109.

27. This is indeed the procedure of Galileo, cf. Chapter 12, footnote 4.

Christology.[28] And so the lever is strengthened further until it uproots the entire orthodox view including its implications about the position of man in the material universe, the relation between man and God, and so on.[29]

As a third example take an anarchist who is interested in the improvement of *scientific* astronomy only and who views increase of content as a necessary condition of such improvement. He may have convinced himself that increase of content can only be achieved with observations of an entirely new kind and he may start the development by claiming to possess such observations though there is not an ounce of argument to establish the claim. Building content-increase on the new observations entirely, he must reject the old observations and he buries them without ever explaining why they should not be used, thus creating the 'epistemological illusion' described in Chapter 15. The new observations are accepted, the old observations are forgotten and no reasons are ever given for the exchange: reasons do not exist when the change occurs, and they are of no interest when they finally become available. This is how increase of content is *manufactured* by the combined use of enthusiasm, forgetfulness, and historical change.

The last two examples, which are only slightly bowdlerized versions of actual historical developments,[30] establish one point (already made in Chapter 1 above): given any aim, even the most narrowly 'scientific' one, the non-method of the anarchist has a greater chance of succeeding than any well-defined set of standards, rules, prescriptions.[31] (It is only *within* the framework of a fairly comprehensive world view that special rules can be justified, and have a chance of success.) The first example makes it plausible that argument, judiciously used, could have prevented the rise of modern science. Argument may retard science while deception is necessary for advancing it. Add to this what we have learned about the

28. *The Quest for the Historical Jesus*, New York, 1962, p. 5.

29. In this domain there were further ideas and attitudes that could be used to strengthen the Copernican ideology. Cf. Hans Blumenberg, *Die Kopernikanische Wende*, Frankfurt, 1965, as well as I. Seznec, *The Survival of the Pagan Gods*, Princeton, 1963, especially p. 60.

30. Cf. the more detailed account in Chapters 6–12 above.

31. Note that the 'epistemological illusion' that often makes progress possible is not supposed to occur according to Lakatos: 'The scores of the rival sides . . . must be recorded and publicly displayed at all times.' 'History', p. 101; italics in the original.

ordering principles of myth, religious enthusiasm, abnormal experiences, and one will be strongly inclined to believe that there are many different ways of approaching nature and society and many different ways of evaluating the results of a particular approach, that we must make a choice, and that there are no *objective* conditions to guide us. So far, a brief and very incomplete sketch of the ideology of epistemological anarchism and of some possible applications.

Imre Lakatos, on the other hand, wants science and, indeed, the whole of intellectual life to conform to certain fixed standards, he wants it to be 'rational'. This means two things: (a) The chosen standards must never be overruled by standards of a different kind; if knowledge, or science, is to be part of a larger context, then this must not affect its nature; science especially must retain its 'integrity'. (b) The standards must have heuristic force as well, that is, the activity that is governed by them must be different from the intellectual freelancing of the anarchist.

Now we have seen that the particular standards which Lakatos has chosen neither issue abstract orders (such as 'eliminate theories that are inconsistent with accepted basic statements') nor do they contain general judgements concerning the rationality or irrationality of a course of action (such as 'it is irrational to stick to a theory that contradicts accepted basic statements'). Such orders and such judgements have given way to concrete decisions in complex historical situations. If the enterprise that contains the standards is to be different from the 'chaos' of anarchism, *then these decisions must be made to occur with a certain regularity*. The standards by themselves cannot achieve this, as we have seen. But psychological or sociological *pressures* may do the trick.

Thus assume that the institutions which publicize the work and the results of the individual scientist, which provide him with an intellectual home where he can feel safe and wanted and which because of their eminence and their (intellectual, financial, political) pull can make him seem important adopt a *conservative attitude* towards the standards, they refuse to support degenerating research programmes, they withdraw money from them, they ridicule their defenders, they refuse to publish their results, they make them feel bad in any way possible. The outcome can be easily foreseen: scientists who are as much in need of emotional and financial support as everyone else, especially today, when science has ceased to be a philosophical adventure and has become a business, will revise their 'decisions' and they will tend to reject research programmes on a downward trend.

Now, the conservative attitude adopted by the institutions is not irrational, for it does not conflict with the standards. It is the result of collective policies of the kind encouraged by the standards. The attitude of the individual scientist who adapts so readily to pressures is not irrational either, for he again decides in a way that is condoned by the standards. We have therefore achieved law and order without reducing the liberalism of our methodology. And even the complex nature of the standards now receives a function. For while the standards do not prescribe, or forbid, any particular action, while they are perfectly compatible with the 'anything goes' of the anarchist who is therefore right in regarding them as mere embroideries, they yet give content to the actions of individuals and institutions who have decided to adopt a conservative attitude towards them. *Taken by themselves*, the standards are incapable of forbidding the most outrageous behaviour. *Taken in conjunction with* the kind of *conservatism* just described they have a subtle but firm influence on the scientist. *And this is precisely how Lakatos wants to see them used:* considering a degenerating programme he suggests that 'editors of scientific journals should refuse to publish . . . papers [by scientists pursuing the programme]. . . . Research foundations, too, should refuse money.'[32] The suggestion is not in conflict with the standards, as we have seen. Given the standards as a measure of rationality it is perfectly proper to make it, and to act accordingly. It puts teeth into the standards not by strengthening their power in argument, but by creating a historical situation in which it becomes very difficult, *practically*, to pursue a degenerating research programme. A research programme is now dropped not because there are arguments against it, on the basis of the standards, but because its defenders cannot go on. Briefly, but not at all unjustly: research programmes disappear not because they get killed in argument but because their defenders get killed in the struggle for survival. It may *seem* that a kindly colleague who expounds on the comparative merits of two research programmes, who gives a detailed account of the success of the one and of the increasing

32. 'History', p. 105.

number of failures of the other, who describes all the *ad hoc* measures, the inconsistencies, the empty verbalism of the degenerating programme, is using very powerful *arguments* against its retention – but such an impression occurs only if one has not yet made the move from naive falsificationism, etc., etc., to Lakatos. A person who *has* made this move and who is aware of the implications of his newly adopted rationality can always reply: 'My dear fellow. You mean well, but you are not up-to-date as far as your theory of rationality is concerned. You think you can convince me by your arguments while I know that in my sense of "rational" one may rationally stick to a degenerating research programme until it is overtaken by a rival, *and even after*.[33] Of course, you may be under the impression that in addition to having accepted the standards of Lakatos, I have also adopted a conservative attitude towards them. Were this the case, then your argument would justly chide me for first making a decision and then not living up to it. But I am not a conservative, I have never been, and so you can force me out of the game, but you cannot show that I have been irrational.'

To sum up: in so far as the methodology of research programmes is 'rational', it does not differ from anarchism. In so far as it differs from anarchism, it is not 'rational'. Even a complete and unquestioning acceptance of this methodology does not create any problem for an anarchist who certainly does not deny that methodological rules may be and usually are enforced by threats, intimidation, deception. This, after all, is one of the reasons why he mobilizes (not counter-arguments but) counter-*forces* to overcome the restrictions imposed by the rules.

It is also clear that Lakatos has not succeeded in showing 'rational change' where 'Kuhn and Feyerabend see irrational change'.[34] My own case has just been discussed. As regards Kuhn, we need only remember that a revolution occurs whenever a new research programme has accumulated a sufficient number of successes and the orthodox programme suffered a sufficient number of failures for both to be considered as serious rivals, and when the protagonists of the new programme proclaim the demise of the orthodox view. Seen from the point of view of the methodology of research programmes they do this not just because of their standards, but because they have adopted a conservative attitude

33. ibid., p. 104.

34. ibid., p. 118; cf. 'Falsification', p. 93.

towards their standards. Their orthodox opponents have what one might call a 'liberal' attitude, they are prepared to tolerate a lot more degeneration than the conservatives. The standards permit both attitudes. They have nothing to say about the 'rationality' or 'irrationality' of these attitudes, as we have seen. It follows that the fight between the conservatives and the liberals and the final victory of the conservatives is not a 'rational change'[35] but a 'power struggle' pure and simple, full of 'sordid personal controversy'.[36] It is a topic not for methodology, or for the theory of rationality, but for 'mob psychology'.[37]

The failure of Lakatos to keep his promise and to reveal the work of reason where others see just a lot of pushing and pulling remains concealed by his ambiguous terminology. On the one side he tells us that the apparent irrationality of many important scientific developments was due to an unnecessarily narrow idea of what is to be counted as rational. If only the acceptance of *proven* theories is rational, if it is irrational to retain theories that are *in conflict* with accepted basic statements, then all of science is irrational. So Lakatos develops new standards. The new standards which are also new measures of rationality no longer forbid what makes good science. But they do not forbid anything else either. They must be strengthened. They cannot be strengthened by adding further standards, i.e. by making *reason* tougher. But they can be given *practical* force by making them the core of conservative *institutions*. Measured by the standards of the methodology of research programmes this conservatism is neither rational nor irrational. *But it is eminently rational according to other standards*, for example, according to the standards of *common sense*.[38] This wealth of meanings of the word 'rational' is used by Lakatos to maximum effect. In his arguments against naive

35. ibid., p. 118. 36. ibid., p. 120.

37. 'Falsification', p. 178 – italics in original.

38. 'In such decisions', says Lakatos, referring to decisions such as those leading to a conservative use of the standards, 'one has to use one's *common sense*' – 'History', footnote 58. Right on – as long as we recognize that in doing so we *leave* the domain of rationality as defined by the standards and move on to an 'external' medium, or to other standards. Lakatos does not always make the change clear. Quite the contrary. In his attack upon opponents he makes full use of our inclination to regard common sense as inherently rational and to use the word 'rational' in accordance with *its* standards. He accuses his opponents of 'irrationality'. We instinctively agree with him, quite forgetting that his own methodology does not support the judgement and does not provide any reasons for making it. Cf. also the next footnote.

falsificationism he emphasizes the new 'rationalism' of his standards which permits science to survive. In his arguments against Kuhn and against anarchism he emphasizes the entirely different 'rationality' of common sense but without informing his audience of the switch, and so he can have his cake – have more liberal standards – and eat it too – have them used conservatively, and he can even expect to be regarded as a rationalist in both cases. Indeed, there is a great similarity between Lakatos and the early Church Fathers who introduced revolutionary doctrines in the guise of familiar prayers (which formed the common sense of their time) and who thereby gradually transformed common sense itself.[39]

This great talent for ambiguous aggression makes Lakatos a most welcome ally in the fight against Reason. For a view that *sounds* 'rational' *in any sense of this emotionally charged word* has today a much greater chance of being accepted than a view that openly rejects the authority of Reason. Lakatos' philosophy, his anarchism in disguise, is a splendid Trojan horse that can be used to smuggle real, straightforward, 'honest' (a word very dear to Lakatos) anarchism into the minds of our most dedicated rationalists. And once they discover that they have been had they will be much less reluctant to concede that the ideology of rationalism has no intrinsic advantage, they will realize that even in science one is subjected to propaganda and involved in a struggle between opposing forces and they will agree that argument is nothing but a subtle and most effective way of paralysing a trusting opponent.[40]

So far I took Lakatos' standards for granted, I compared them with other standards, I asked how they influence behaviour (for example, I asked how a practice guided by the methodology of research programmes

39. Using the *psychological* hold which the baptismal confession had over the members of the early Christian Churches and taking the non-Gnostic interpretation 'as its self-evident content' (Von Harnack, *History of Dogma*, Vol. II, New York, 1961, p. 26), Irenaeus succeeded in defeating the Gnostic heresy. Using the psychological hold which common sense has over philosophers of science and other creatures of habit and taking the conservative interpretation of his standards as *its* self-evident content, Imre Lakatos has almost succeeded in convincing us of the rationality of his law-and-order philosophy and the non-ornamental character of his standards: now, as before, the best propagandists are found in the Church, and in conservative politics.

40. For some objections which are usually raised at this point, cf. the Appendix to this chapter.

differs from an anarchistic practice), and I examined the implications of the standards for the theory of rationality. Now comes the question why we should consider the standards at all, why we should prefer them to other *scientific* standards such as inductivism, or to '*unscientific*' standards such as the standards of religious fundamentalists. Lakatos gives an answer to the first question but not to the second, though he succeeds in creating the impression that he has answered them both. Here as before he uses common sense and the general predilection for science to help him across chasms he cannot bridge by argument. Let us see how he proceeds!

I have said that both Lakatos and I evaluate methodologies by comparing them with historical data. The historical data which Lakatos uses are ' "basic" appraisals of the scientific elite'[41] or 'basic value judgements'[42] which are *value* judgements about *specific* achievements of science. Example: 'Einstein's theory of relativity of 1919 is superior to Newton's celestial mechanics in the form in which it occurs in Laplace.' For Lakatos such value judgements (which together form what he calls a 'common scientific wisdom') are a suitable basis for methodological discussions because they are accepted by the great majority of scientists: 'While there has been little agreement concerning a *universal* criterion of the scientific character of theories, there has been considerable agreement over the last two centuries concerning *single* achievements.'[43] Basic value judgements can therefore be used for checking theories about science or *rational reconstructions* of science much in the same way in which 'basic' *statements* are used for checking theories about the world. The ways of checking depend of course on the particular methodology one has chosen to adopt: a falsificationist will reject methodological rules *inconsistent* with basic value judgements,[44] a follower of Lakatos will accept methodological research programmes which 'represent a *progressive shift* in the sequence of research programmes of rational reconstructions: . . . progress in the theory of scientific rationality is marked by discoveries of novel historical facts, by the reconstruction of a growing bulk of value impregnated history as rational'.[45] The standard of methodological criticism thus turns out to be the best methodological research

41. 'History', p. 111.
42. ibid., p. 117.
43. ibid., p. 111.
44. Cf. the rule in 'History', p. 111.
45. 'History', pp. 117–18.

programme that is available at a particular time. So far a first approximation of the procedure of Lakatos.

The approximation has omitted two important features of science. On the one side basic value judgements are not as uniform as has been assumed. Science is split into numerous disciplines, each of which may adopt a different attitude towards a given theory and single disciplines are further split into schools. The basic value judgements of an experimentalist will differ from those of a theoretician (just read Rutherford, or Michelson or Ehrenhaft on Einstein), a biologist will look at a theory differently from a cosmologist, the faithful Bohrian will regard modifications of the quantum theory with different eyes than will the faithful Einsteinian. Whatever unity remains is dissolved during revolutions, when no principle remains unchallenged, no method unviolated. Even individual scientists arrive at different judgements about a proposed theory: Lorentz, Poincaré, Ehrenfest thought that Kaufmann's experiments had refuted the special theory of relativity and were prepared to abandon the relativity principle in the form proposed by Einstein while Einstein himself was of a different opinion.[46] Secondly, basic value judgements are only rarely made for good reasons. Everyone agrees that *Copernicus' hypothesis* was a big step forward, but hardly anyone can give a halfway decent account of it,[47] let alone enumerate the reasons for its excellence. *Newton's theory* (of gravitation) was 'highly regarded by the greatest scientists',[48] most of whom were unaware of its difficulties and some of whom believed that it could be derived from Kepler's laws.[49] The *quantum theory*, which suffers from quantitative and qualitative disagreements with the evidence[50] and is also quite clumsy in places, is accepted not *despite* its difficulties, in a *conscious violation* of naive falsificationism, but because 'all evidence points with merciless definiteness in the . . . direction . . . [that] all the processes involving . . .

46. For literature cf. footnotes 32 and 33 of my essay 'Von der beschränkten Gültigkeit methodologischer Regeln', *Neue Hefte für Philosophie*, Heft 2/3, Göttingen, 1972, as well as footnotes 6 and 9 of Chapter 5.

47. Cf. the brief survey on pp. 139ff of 'Von der, etc.' as well as Chapters 6–12 of this essay.

48. 'History', p. 112.

49. M. Born, *Natural Philosophy of Cause and Chance*, London, 1948, pp. 129ff.

50. Cf. footnotes 5 and 17–19 of Chapter 5.

unknown interactions conform to the fundamental quantum law'.[51] And so on. *These* are the reasons which produce the basic value judgements whose 'common scientific wisdom' Lakatos occasionally gives such great weight.[52] Add to this the fact that most scientists accept basic value judgements on trust, they do not examine them, they simply bow to the authority of their specialist colleagues and you will see that *'common scientific wisdom' is not very common and it certainly is not very wise.*

Lakatos is aware of the difficulty. He realizes that basic value judgements are not always reasonable,[53] and he admits that 'the scientists' judgement [occasionally] fails'.[54] In such cases, he says, it is to be balanced and perhaps even overruled, by the 'philosopher's statute law'.[55] The 'rational reconstruction of science' which Lakatos uses as a measure of method is therefore not just the sum total of all basic value judgements, nor is it the best research programme trying to absorb them. It is a 'pluralistic system of authorities'[56] in which basic value judgements are a dominating influence as long as they are uniform *and* reasonable. But when the uniformity disappears, or when 'a tradition degenerates',[57] then general philosophical constraints come to the fore and enforce (restore) reason and uniformity.

Now I have the suspicion that Lakatos vastly underestimates the number of occasions when this is going to be the case. He believes that uniformity of basic value judgements prevailed 'over the last two centuries'[58] when it was actually a very rare event. But if that is the case then his 'rational reconstructions' are dominated either by common sense[59] or by the abstract standards and the concrete pressures of the methodology of research programmes. Moreover, he accepts a uniformity only if it does not stray too much from his standards: 'When a scientific school degenerates into pseudoscience, it may be worthwhile to force a methodological debate.'[60] This means that the judgements which Lakatos passes so freely are ultimately neither the results of research nor parts of

51. Rosenfeld in *Observation and Interpretation*, London, 1957, p. 44.

52. 'Is it not . . . *hubris* to try to impose some *a priori* philosophy of science on the most advanced sciences? . . . I think it is.' 'History', p. 121.

53. ibid., footnote 80.

54. ibid., p. 121.

55. ibid., p. 121.

56. ibid., p. 121.

57. ibid., p. 122.

58. ibid., p. 111.

59. Cf. footnote 38 above.

60. 'History', p. 122.

'scientific practice'; they are parts of an *ideology* which he tries to impose on us in the guise of a ' "common" scientific wisdom'. For a second time we encounter a most interesting difference between the *wording* of Lakatos' proposals and their *cash value*. We have seen that the methodology of research programmes was introduced with the purpose of aiding rationalism. Yet it cannot condemn a single action as 'irrational'. Whenever Lakatos makes *such* a judgement – and he does this often enough – he relies on 'external' agencies, for example he relies on his own conservative inclinations or on the conservatism inherent in common sense. Now we discover that his 'reconstructions' are much closer to the general methodologies he claims he is examining and that they merge with them in times of crisis. Despite the difference of rhetorics ('Is it not . . . *hubris* to try to impose some *a priori* philosophy of science on the most advanced sciences? . . . I think it is'[61]), despite the decision to keep things concrete ('there has been considerable agreement . . . concerning *single* achievements'[62]), Lakatos does not really differ from the traditional epistemologists; quite the contrary, he provides them with a powerful new propaganda device: he connects his principles with what at first seems like a substantial bulk of independent scientific common sense, but this bulk is neither very substantial nor independent. It is shot through and constituted in accordance with the abstract principles he wants to defend.

Let us look at the matter from a different point of view. A 'rational reconstruction' in the sense of Lakatos comprises concrete judgements about results in a certain domain, as well as general standards. It is 'rational' in the sense that it reflects *what is believed to be a valuable achievement* in the domain. It reflects what one might call the *professional ideology* of the domain. Now even if this professional ideology consisted of a uniform bulk of basic value judgements only, even if it had no abstract ingredients whatsoever, even then *it would not guarantee that the corresponding field has worthwhile results, or that the results are not illusory*. Every medicine-man proceeds in accordance with complex rules, he compares his results and his tricks with the results and the tricks of other medicine-men of the same tribe, he has a rich and coherent professional ideology – and yet no rationalist would be inclined to take

61. ibid., p. 121. 62. ibid., p. 111.

him seriously. Astrological medicine employs strict standards and contains fairly uniform basic value judgements, and yet rationalists reject its entire professional ideology as 'irrational'. For example, they are not prepared even to consider the 'basic value judgement' that the tropical method of preparing a chart is preferable to the sidereal method (or vice versa[63]). This possibility to reject professional standards *tout court* shows that 'rational reconstructions' *alone* cannot solve the problem of method. To find the right method, one must reconstruct the *right discipline*. But what is the right discipline?

Lakatos does not consider this question – and there is no need for him to do so as long as he merely wants to know what is going on in post 17th-century science and as long as he can take it for granted that this enterprise rests on a coherent and uniform professional ideology. (We have seen that it does not.) But Lakatos goes further. Having finished his 'reconstruction' of modern science, he turns it against other fields *as if it had already been established* that modern science is superior to magic, or to Aristotelian science, and that it has no illusory results. However, there is not a shred of an argument of this kind. 'Rational reconstructions' take 'basic scientific wisdom' *for granted*, they do not *show* that it is better than the 'basic wisdom' of witches and warlocks. Nobody has shown that science (of 'the last two centuries'[64]) has results that conform to its own 'wisdom' while other fields have no such results. What *has* been shown, by more recent anthropological studies is that *all* sorts of ideologies and associated institutions produce, and have produced, results that conform to their standards and other results which do not conform to their standards. For example, Aristotelian science has been able to accommodate numerous facts without changing its basic notions and its basic principles, thus conforming to its own standard of *stability*. We obviously need further considerations for deciding which field to accept as a measure of method.

Exactly the same problem arises in the case of *individual* methodological rules. It is hardly satisfactory to reject naive falsificationism because it conflicts with some basic value judgements of eminent scientists. Most of these eminent scientists retain refuted theories not because

63. 'Vice versa' – this was Kepler's opinion. Cf. Norbert Herz, *Keplers Astrologie*, Vienna, 1895, and the references given there.

64. 'History', p. 111.

they have some insight into the limits of naive falsificationism, but because they do not realize that the theories are refuted (cf. the examples in text to footnotes 46–50 of the present chapter). Besides, even a more 'reasonable' practice would not be sufficient to reject the rule: universal leniency towards refuted theories may be nothing but a mistake. It certainly is a mistake in a world that contains well-defined species which are only rarely misread by the senses. In such a world the basic laws are manifest and recalcitrant observations are rightly regarded as indicating an error in our *theories* rather than in our *methodology*. The situation changes when the disturbances become more insistent and assume the character of an everyday affair. A cosmological discovery of this kind forces us to make a choice: shall we retain naive falsificationism and conclude that knowledge is impossible; or shall we opt for a more abstract and recondite idea of knowledge and a correspondingly more liberal (and less 'empirical') type of methodology? Most scientists, unaware of the nomological-cosmological background of the problem and even of the problem itself, retain theories that are incompatible with established observations and experiments and praise them for their excellence. One might say that they make the right choice *by instinct*[65] – but one will hardly regard the resulting behaviour as a decisive measure of method, especially in view of the fact that the instinct has gone wrong on more occasions than one. The *cosmological criticism* just outlined (omnipresence of disturbances) is to be preferred.

A cosmological criticism[66] gains in importance when new methods and new forms of knowledge appear on the scene. In periods of degeneration, says Lakatos, the philosopher's statute law comes to the fore and tries to 'thwart the authority of the corrupted case law' of the scientist.[67] Examples of incipient or protracted degeneration he has in mind are certain parts of sociology, social astrology,[68] modern particle physics.[69]

65. 'Up to the present day it has been the scientific standards, as applied "instinctively" by the scientific *elite* in *particular cases* which have constituted the main – although not the exclusive – yardstick of the philosopher's *universal* laws.' 'History', p. 121.

66. 'Cosmology' here comprises history, sociology, psychology and all other factors that may influence the success of a certain procedure. The 'law' of uneven development which I mentioned in Chapter 12 also belongs to 'cosmology' in that sense.

67. 'History', p. 122.

68. ibid., footnote 132; 'Falsification', p. 176.

69. 'History', footnote 130.

All these cases violate 'good methodology'[70] which is a methodology ' "distilled" from mature science'[71]; in other words, they violate the professional ideology of the science of Newton, Maxwell, Einstein (though not of Bohr[72]). But the restless change of modern science that announces itself with Galileo, its loose use of concepts, its refusal to accept customary norms, its 'unempirical' procedures violated the professional ideology of the Aristotelians and was an example of incipient degeneration *for them.* In forming this judgement the Aristotelians made use of *their* general philosophy, *their* desiderata (creation of a stable intellectual order based on the same type of perception that helps man in his everyday affairs, 'saving of phenomena' with the help of mathematical artifices, etc.), and of the basic value judgements of *their* science (which disregarded the Occamists just as Lakatos now disregards the Copenhagen gang). And the Aristotelians had a tremendous advantage, for the basic value judgements of the followers of the Copernican Creed were even more varied and unreasonable than the basic value judgements of elementary-particle physicists are today. In addition, the Aristotelian philosophy was supported by the widespread belief, still found in Newton, that most innovations were of minor importance and that all the important things had already been found. It is clear that a 17th-century Lakatos would have sided with the schools. *And thus he would have made the same 'wrong' decisions as a 17th-century inductivist, or a 17th-century conventionalist, or a 17th-century falsificationist.* We see again that Lakatos has not overcome the difficulty which cataclysmic developments in the sciences pose for other methodologies; he has not succeeded in showing that such developments can be seen in their entirety through 'Popperian spectacles'.[73] Once more a methodologist is forced to admit that the quarrel between the Ancients and the Moderns cannot be reconstructed in a rational way. At least, such a reconstruction cannot be given *at the time of the quarrel.*

70. ibid., footnote 132. 71. ibid., p. 122.

72. ibid., footnote 130; 'Falsification', p. 145: 'The *rational position* is best characterized by Newton [ ]'. We see how *arbitrary* this selection of standards is: the lonely Einstein is accepted, the well-disciplined cohorts of the Copenhagen school are pushed aside. One certainly does not need the whole complicated machinery of basic value statements, balanced by 'common sense' and philosophical principles when one knows in advance what developments one is not going to tolerate.

73. 'Falsification', p. 177.

Yet today the situation is exactly the same. It is of course possible to 'reconstruct' the transition by replacing the Aristotelian basic value judgements (about Aristotelian theories) by modern basic value judgements and to use modern standards (progress with increase of content) instead of the Aristotelian standards (stability of principles; *post hoc* 'saving of phenomena'). But first of all the need for *such* a 'reconstruction' would show what Lakatos denies, viz. that 'new paradigms bring [in] a . . . new rationality'.[74] And secondly, one would have rejected the professional ideology of the Aristotelians without having shown that it is worse than its replacement: in order to decide between a 'rational reconstruction' (in Lakatos' sense) of Aristotelian science that utilizes the 'statute law' of the Aristotelian philosophy and the basic value judgements of the best Aristotelian scientists, and a 'rational reconstruction' of 'modern' science (of the 'last two centuries'[75]) based on 'modern' statute law and 'modern' basic value judgements, one needs more than 'modern' standards and 'modern' basic value judgements. One must either show that at the time in question the Aristotelian methods did not reach the Aristotelian aims or that they had great difficulties reaching them while the 'moderns', using modern methods, experienced no such difficulties relative to *their* aims, *or* one must show that the modern aims are preferable to the Aristotelian

74. ibid., p. 178.

75. 'History', p. 111. All the methodological judgements of Lakatos are based (if they are based on basic statements at all – see text to footnotes 58ff of the present chapter) on the basic value judgements and the statute law of this period disregarding the basic value judgements of schools he does not like. And when the basic value judgements do not show the required unity, then they are at once replaced by Popperian standards. Small wonder that Lakatos does not find even a trace of 'scientific knowledge' in the Middle Ages. For at that time thinkers did indeed proceed in a very different fashion. Using his standards Lakatos cannot say that they were worse – and so he simply falls back on the vulgar ideology of our own 'scientific' age. Most research into Egyptian, Babylonian, and Ancient Greek astronomy proceeds in exactly the same way. It is interested only in those fragments of the older ideas which conform to the ideology of modern science. It disregards the older cosmologies and the older aims that united them and other fragments in a most impressive way. Small wonder the results look incoherent and 'irrational'. A lonely exception is B. L. van der Waerden, *Erwachende Wissenschaft*, II, Basle, 1968, p. 7: 'In this book we examine the history of Babylonian astronomy in its interaction with the astral religion and astrology. Using this method we do not tear astronomy out of the cultural-historical pattern to which it belongs.' Cf. also my *Einführung in die Naturphilosophie*, Braunschweig, 1976, where the transition from myth to philosophy is examined in some detail.

aims. Now we have seen that the 'Aristotelians'[76] were doing very well while the 'moderns' were faced by numerous problems which they concealed with the help of propagandistic stratagems.[77] If we want to know why the transition occurred and how it can be justified given *our* predilection for the methods and results of contemporary science, then we will have to identify the *motives* that made people proceed despite the problems,[78] and we will also have to examine the *function* of propaganda, prejudice, concealment and of other 'irrational' moves in the gradual solution of the problems. All these are 'external' factors in the scheme of Lakatos.[79] Yet without them there is no way of understanding one of the major revolutions of thought. Without them we can only say that the professional ideology of 15th- and 16th-century physics and astronomy was followed by the professional ideology of 'modern' science and that the latter now reigns supreme. We cannot explain how this came about, nor do we have any reason for asserting that our professional ideology is better than that of the Aristotelians.

Let me now give a short, incomplete, and very one-sided sketch of the transition that considers factors I think are relevant and explains their function in the rise of the new astronomy. Many details are missing while others are exaggerated. But my purpose is not to provide a scholarly account, my purpose is to tell a *fairy-tale* that might some day become a scholarly account and that is more realistic and more complete than the

76. I repeat that I do not here refer to the doctrines contained in the Aristotelian corpus but to their elaboration within the astronomy, psychology, etc., etc., of the later Middle Ages. The term 'Aristotelians' is a simplification, of course, and must some day be replaced by an account of the influence of individual thinkers. In the meantime we can use it in our criticism of another simplification, viz. 'modern' science 'of the last two centuries'.

77. They are sheer propaganda when judged by Lakatos' standards. Realization of the function they have in the rise of modern science improves our opinion of them and so we call them 'rational'.

78. In many respects the relation between the Aristotelians and the followers of Copernicus is comparable to the relation between the members of the Copenhagen School and the hidden-variable theoreticians. The ones establish basic principles and then give a purely formal account of newly discovered facts while the others want the basic principles themselves to anticipate and/or to explain all the relevant facts. Considering the difficulties of any unified account the first method would seem to be considerably more realistic.

79. 'History', section i/E.

fairy-tale insinuated by Lakatos and his mafia. For details the reader is advised to return to Chapters 6–12 of the present essay.

To start with, we must admit that new basic value judgements and a new statute law now enter astronomy. There are not only new theories, new facts, new instruments, *there is also a new professional ideology*.[80] The ideology is not invented out of the blue, it has its ancestors in antiquity (Xenophanes, Democritos, for example) and plays some role in trades and professions outside physics and astronomy. The rising importance of the classes and groups engaged in these trades and professions makes the ideology important as well and lends support to those who want to use it inside astronomy. Their support is urgently needed, for the many theoretical difficulties that arise can be solved only if one has sufficient determination to go ahead with the programme of the moving earth. The different emphasis given by the new classes to 'Copernicus' (progress, forward looking, against the *status quo*) and to 'Aristotle' (backward looking, for the *status quo*, hostile to the emergence of the new classes) increases the determination, reduces the impact of the difficulties and makes astronomical progress possible. This association of astronomical ideas and historical (and class) tendencies does not make the Aristotelian arguments less rational or less conclusive, but it reduces their influence on the minds of those who are willing to follow Copernicus. Nor does it produce a single new argument. But it engenders a firm commitment to the idea of the motion of the earth – and that is all that is needed at this stage, as we have seen. (We have also seen in earlier chapters, how masterfully Galileo exploits the situation and how he amplifies it by tricks, jokes, *non sequiturs* of his own.) Which brings me to the second point.

Our problem is: given the historical situation of the idea of the moving earth in, say, 1550 and its historical situation in, say, 1850 – how was it possible to get from the first situation (S′) to the second (S″)? What

80. This is overlooked in Lakatos–Zahar's forthcoming 'Did Copernicus supersede Ptolemy?', where it is assumed that the methodology used for evaluating theories (1) remained unchanged throughout the transition from Ptolemy to Copernicus and (2) did not differ significantly from the methodology of research programmes. (Lakatos and Zahar also disregard the difficulties of dynamics which have been discussed in Chapters 6 and 7 above. Adding these difficulties to their success story turns it into a story of dismal failure.)

psychological, historical, methodological conditions had to be satisfied so that a group of people dedicated to the improvement of knowledge, and especially of astronomy, could move science – and this means the professional prejudices of the astronomers as well as the conditions outside science that are necessary for its survival in a particular form – from S′ to S″? Conversely, what beliefs, actions, attitudes would have made it impossible to reach S″ from S′? We see at once that the arrival of a new professional ideology was absolutely essential – but this is a point not accessible to analysis in the terms provided by Lakatos. We also see that the distinction between 'internal' and 'external' history that is so important for Lakatos restricts the answer and protects the methodology that was chosen as its basis. *For it is quite possible that a science has a certain 'internal' history only because its 'external' history contains compensating actions that violate the defining methodology at every turn.* Examples can be readily provided. Galileo's ignorance of the basic principles of telescopic vision will certainly be put into the external part of the history of astronomy. But given S′, that is, given 16th-century optical and psychological theories, this ignorance was necessary for Galileo to speak as decisively as he did. In the historical situation this ignorance was bliss. His as yet unsupported belief in Copernicanism was necessary for interpreting what he saw *as evidence*, and, more specifically, as evidence for the essential similarity between things above and things below. The existence of groups of anti-Aristotelians and of other enemies of school philosophers was necessary for turning such subjective acts into a more comprehensive social phenomenon and, finally, into the elements of a new science. Concentrating on the internal history of Copernicanism we notice an increase of content (Galileo's observations), and so we seem to be in agreement with the principles of the new professional ideology. But adding the external history or, as Lakatos expresses himself, adding 'mob psychology' to our information we notice *that the agreement 'inside' science is the result of numerous violations 'outside' of it*, we realize that these violations were necessary for the transition from S′ to S″ *and that they therefore belong to science itself*, and not to some other domain. For example, the increase of content which Lakatos regards with such pride is a result of the 'epistemological illusion' I described earlier, which in turn arises only because one decides *not* to 'record' and to 'publicly display' the true 'scores of the rival sides'.[81] Thus even a development that looks quite orderly must be constantly checked, which means that the separation between 'internal' and 'external' (and the corresponding separation between a Third Heaven and its disorderly reflection in human minds[82]) inhibits the study of scientific change. It is just another example of a distinction without a difference which, if taken seriously, will make a lot of difference to the quality of our research.

Finally, there is some doubt whether Lakatos' criterion of content-increase which plays such an important role in his standards satisfies his own conditions for an acceptable rationality theory (see above, footnote 12 and text to footnotes 41ff; note that I am here not considering the problem of incommensurability!). Considering the omnipresence of the 'epistemological illusion' and the development of such research programmes as atomism, the moving earth, physicalism in the sense that the world at large obeys the laws of physics without any divine interference, we may have to conclude that increase of content (compared with the content of rival programmes) *is an extremely rare event* and that the historical research programme that assumes its existence has been degenerating and is still degenerating. However, there is not yet sufficient evidence available for making this conclusion palatable to an empiricist.

I thus arrive at the following evaluation of Lakatos' achievement.

All theories of (scientific) knowledge proceed from the question: what is knowledge, and how can it be obtained?

The *traditional answer*[83] contains a definition of knowledge, or potential knowledge (a criterion of demarcation), and an enumeration of the steps by means of which knowledge can be obtained (by means of which knowledge can be separated from non-knowledge). The traditional answer is usually regarded as final. At any rate, one only rarely learns how it can be revised.[84] The revisions that do occur are surreptitious,

81. For the 'epistemological illusion' cf. Chapter 15, text to footnotes 6 and 7. The quotation is from 'History', p. 101.

82. 'Falsification', p. 180; 'History', section i/E.

83. This way of speaking is of course a simplification. So is the description that follows.

84. This is true of Popper: 'He does not raise, let alone answer the questions: "*Under what circumstances would you give up your demarcation criterion?*",' 'History', p. 110,

unaccompanied by argument, they often change the knowledge-gathering practice but without changing the accompanying epistemology.[85] As a result the contact between science and epistemology becomes more tenuous and finally disappears altogether.[86] This is the situation I have described in the preceding chapters of this essay.[87] Nobody admits that there could be various forms of knowledge and that it might be necessary to make a choice.

Compared with this traditional theory the *theory of Lakatos* is a tremendous improvement. His standards and his conception of knowledge are much closer to science than those of the preceding accounts, they can be revised, or so it seems, and we also learn how to carry out the revision. The methods of revision involve history in an essential way and thus close the gap between the *theory* of knowledge and the material (the 'knowledge') that is *actually* being assembled. It is now possible to discuss even the simplest rule in a realistic way and to decide whether it should be retained or replaced by a different rule. This is the impression created by the way in which Lakatos *presents* his methodology, this is how it *appears* to the unwary and enthusiastic reader. A closer look, a more 'rational' examination, reveals an entirely different story: Lakatos has not shown that his standards are the standards of science, he has not shown that they lead to substantial results, he has not even succeeded in

---

italics in the original. It does *not* apply to Plato or Aristotle, who *examine* knowledge and *discover* its complexity. Cf. W. Wieland, *Die Aristotelische Physik*, pp. 76ff. (The whole fuss which Popperians make about 'background knowledge' is here anticipated with strong and simple arguments and observations.) But it *does* apply to the Aristotelians of the later Middle Ages.

85. An example is described in my 'Classical Empiricism', *The Methodological Heritage of Newton*, ed. Butts, Oxford, 1969.

86. As an example cf. the relation between Descartes' philosophy and his physics, between Newton's methodology and his physics, and between Popper's philosophy and Einstein's physics *as seen by Einstein*. The last case is somewhat obscured by the fact that Popper mentions Einstein as one of the inspirations and as the main instance of his own falsificationism. Now it is quite possible that Einstein, who seems to have been somewhat of an epistemological opportunist (or cynic – cf. text to footnote 6 of the Introduction), occasionally said things that can be construed as supporting a falsificationist epistemology. However his actions and the bulk of his written utterances tell a different story. Cf. Chapter 5, footnote 9.

87. Cf. also my talk at the German Conference of Philosophy, Kiel, October 1972, to appear in the *Proceedings* (Felix Mainer, Hamburg).

giving them force except by using pressure, intimidation, threats. He has not refuted anarchism, nor has he even shown that his methodology is the better historical research programme. He arbitrarily selects science as a measure of method and knowledge without having examined the merits of other professional ideologies. For him these ideologies are simply non-existent. Disregarding them he gives only a caricature of major social and intellectual upheavals, disregarding 'external' influences he falsifies the history of subjects by insinuating that deviations from the standards were not *necessary* for their progress. This is the 'True Story' of Imre Lakatos. However, as I have said, *this is not the story that influences the reader*. As in other cases the student of the methodology of research programmes is influenced by its appearance, *not* by its 'rational' core ('rational' now understood in the sense of the theory of rationality defended by Lakatos). And as this appearance represents a tremendous step beyond even the reality of earlier views, as it has led to most interesting historical and philosophical discoveries, and as it seems to provide a clear, unambiguous guide through the maze of history, we can support it without giving up anarchism. We may even admit that at the present stage of philosophical consciousness an irrational theory falsely interpreted as a new account of Reason will be a better instrument for freeing the mind than an out-and-out anarchism that is liable to paralyse the brains of almost everyone. (Having concluded my essay, I shall therefore join Lakatos rather than continuing to beat the drum of *explicit* anarchism.) On the other hand, there is no reason why one should not try to anticipate the next stage by assembling obstacles and presenting them in as impressive a manner as possible. Let us therefore take a look at the phenomenon of *incommensurability* which to my mind creates problems for all theories of rationality, including the methodology of research programmes. The methodology of research programmes assumes that rival theories and rival research programmes can always be compared with respect to their content. The phenomenon of incommensurability seems to imply that this is not the case. How can this phenomenon be identified, and what are the reasons for its existence?

Appendix 3

Having listened to one of my anarchistic sermons, Professor Wigner replied: 'But surely, you do not read all the manuscripts which people send you, but you throw most of them into the wastepaper basket.' I most certainly do. 'Anything goes' does not mean that I shall read every single paper that has been written – God forbid! – it means that I make my selection in a highly individual and idiosyncratic way, partly because I can't be bothered to read what doesn't interest me – and my interests change from week to week and day to day – partly because I am convinced that Mankind, and even Science, will profit from everyone doing his own thing: a physicist might prefer a sloppy and partly incomprehensible paper full of mistakes to a crystal-clear exposition because it is a natural extension of his own, still rather disorganized, research and he might achieve success as well as clarity long before his rival who has vowed never to read a single woolly line (one of the assets of the Copenhagen School was its ability to avoid premature precision: cf. 'On a Recent Critique of Complementarity, Part II', *Philosophy of Science*, March 1969, sec. 6ff). On some other occasions he might look for the most perfect proof of a principle he is about to use in order not to be sidetracked in the debate of what he considers to be his main results. There are of course so-called 'thinkers' who subdivide their mail in exactly the same way, come rain, come sunshine, and who also imitate each other's principles of choice – but we shall hardly admire them for their uniformity, and we shall certainly not think their behaviour 'rational': Science needs people who are adaptable and inventive, not rigid imitators of 'established' behavioural patterns.

In the case of institutions and organizations such as the National Science Foundation the situation is exactly the same. The physiognomy of an organization and its efficiency depends on its members and it improves with their mental and emotional agility. Even Procter and Gamble have by now realized that a bunch of yes-men is inferior in competitive potential to a group of people with unusual opinions and business has found ways of incorporating the most amazing nonconformists into their machinery. Special problems arise with foundations that distribute money and want to do this in a just and reasonable way. Justice seems to demand that the allocation of funds be carried out on the basis of standards which do not change from one applicant to the next and which reflect the intellectual situation in the fields to be supported. The demand can be satisfied in an *ad hoc* manner without appeal to *universal* 'standards of rationality'. One may even maintain the illusion that the chosen rules guarantee efficiency and are not simply expedient stop-gap measures: any free association of people must respect the illusions of its members and must give them institutional support. The illusion of *rationality* becomes especially strong when a scientific institution opposes political demands. In this case one class of standards is set against another such class – and this is quite legitimate: each organization, each party, each religious group has a right to defend its particular form of life and all the standards it contains. *But scientists go much further*. Like the defenders of the One and True Religion before them they insinuate that their standards are *essential* for arriving at the Truth, or for getting Results and they deny such authority to the demands of the politician. They especially oppose any political interference, and they fall over each other trying to remind the listener, or the reader, of the disastrous outcome of the Lysenko affair.

Now we have seen that the belief in a unique set of standards that has always led to success and will always lead to success is nothing but a chimera. The *theoretical* authority of science is much smaller than it is supposed to be. Its *social* authority, on the other hand, has by now become so overpowering *that political interference is necessary to restore a balanced development*. And to judge the *effects* of such interference one must study more than one unanalysed case. One must remember those cases where science, left to itself, committed grievous blunders and one must not forget the instances when political interference has *improved* the situation (an example was discussed in the text to footnotes 9–13 of Chapter 4). Such a balanced presentation of the evidence may even convince us that the time is overdue for adding the separation of state and science to the by now quite customary separation of state and church.

Science is only *one* of the many instruments man has invented to cope with his surroundings. It is not the only one, it is not infallible, and it has become too powerful, too pushy, and too dangerous to be left on its own.

Finally, a word about the *practical aim* Lakatos wants to realize with the help of his methodology.

Lakatos is concerned about intellectual pollution. I share his concern. Illiterate and incompetent books flood the market, empty verbiage full of strange and esoteric terms claims to express profound insights, 'experts' without brains, without character, and without even a modicum of intellectual, stylistic, emotional temperament tell us about our 'condition' and the means for improving it, and they do not only preach to *us* who might be able to look through them, they are let loose on our children and permitted to drag them down into their own intellectual squalor. 'Teachers' using grades and the fear of failure mould the brains of the young until they have lost every ounce of imagination they might once have possessed. This is a disastrous situation, and one not easily mended. But I do not see how the methodology of Lakatos can help. As far as I am concerned the first and the most pressing problem is to get education out of the hands of the 'professional educators'. The constraints of grades, competition, regular examination must be removed and *we must also separate the process of learning from the preparation for a particular trade*. I grant that business, religions, special professions such as science or prostitution, have a right to demand that their participants and/or practitioners conform to standards they regard as important, and that they should be able to ascertain their competence. I also admit that this implies the need for special types of education that prepare a man or a woman for the corresponding 'examinations'. The standards taught need not be 'rational' or 'reasonable' in any sense, though they will be usually presented as such; it suffices that they are *accepted* by the groups one wants to join, be it now Science, or Big Business, or The One True Religion. After all, in a democracy 'reason' has just as much right to be heard and to be expressed as 'unreason' especially in view of the fact that one man's 'reason' is the other man's insanity. But one thing must be avoided at all costs: the special standards which define special subjects and special professions must not be allowed to permeate *general* education and they must not be made the defining property of a 'well-educated man'. General education should prepare a citizen to *choose between* the standards, or to find his way in a society that contains groups committed to various standards *but it must under no condition bend his mind so that it conforms to the standards of one particular group.* The standards will be *considered*, they will be *discussed*, children will be encouraged to get proficiency in the more important subjects, *but only as one gets proficiency in a game*, that is, without serious commitment and without robbing the mind of its ability to play other games as well. Having been prepared in this way a young person may decide to devote the rest of his life to a particular profession and he may start taking it seriously forthwith. This 'commitment' should be the result of a conscious decision, on the basis of a fairly complete knowledge of alternatives, *and not a foregone conclusion.*

All this means, of course, that we must stop the scientists from taking over education and from teaching as 'fact' and as 'the one true method' whatever the myth of the day happens to be. Agreement with science, decision to work in accordance with the canons of science should be the result of examination and choice, and *not* of a particular way of bringing up children.

It seems to me that such a change in education and, as a result, in perspective will remove a great deal of the intellectual pollution Lakatos deplores. The change of perspective makes it clear that there are many ways of ordering the world that surrounds us, that the hated constraints of one set of standards may be broken by freely accepting standards of a different kind, and that there is no need to reject *all* order and to allow oneself to be reduced to a whining stream of consciousness. A society that is based on a set of well-defined and restrictive rules so that being a man becomes synonymous with obeying these rules, *forces the dissenter into a no-man's-land of no rules at all and thus robs him of his reason and his humanity.* It is the paradox of modern irrationalism that its proponents silently identify rationalism with order and articulate speech and thus see themselves forced to promote stammering and absurdity – many forms of 'mysticism' and 'existentialism' are impossible without a firm but unrealized commitment to some principles of the despised ideology (just remember the 'theory' that poetry is nothing but emotions colourfully expressed). Remove the principles, admit the possibility of many different forms of life, and such phenomena will disappear like a bad dream.

My diagnosis and my suggestions coincide with those of Lakatos – up to a point. Lakatos has identified overly-rigid rationality principles as the source of some versions of irrationalism and he has urged us to adopt new and more liberal standards. I have identified overly-rigid rationality principles as well as a general respect for 'reason' as the source of some forms of mysticism and irrationalism, and I also urge the adoption of more liberal standards. But while Lakatos' great 'respect for great science' ('History', p. 113) makes him look for the standards within the confines of modern science 'of the last two centuries' (p. 111), I recommend to put science in its place as an interesting but by no means exclusive form of knowledge that has many advantages but also many drawbacks: 'Although science taken as a whole is a nuisance, one can still learn from it' (Gottfried Benn, Letter to Gert Micha Simon of 11 October 1949; quoted from Gottfried Benn, *Lyrik und Prosa, Briefe und Dokumente*, Wiesbaden, 1962, p. 235). Also I don't believe that charlatans can be banned just by tightening up rules.

Charlatans have existed at all times and in the most tightly-knit professions. Some of the examples which Lakatos mentions ('Falsification', p. 176, footnote 1) seem to indicate that the problem is created by too much control and not by too little (cf. also his remarks on 'false consciousness' in 'History', pp. 94, 108ff). This is especially true of the new 'revolutionaries' and their 'reform' of the universities. Their fault is that they are Puritans and *not* that they are libertines (for an older example cf. the *Born–Einstein Letters*, New York, 1971, p. 150). Besides, who would expect that cowards will improve the intellectual climate more readily than will libertines? (Einstein saw this problem and he therefore advised people not to connect their research with their profession: research has to be free from the pressures which professions are likely to impose – *Born–Einstein Letters*, pp. 105ff). We must also remember that those rare cases where liberal methodologies *do* encourage empty verbiage and loose thinking ('loose' from one point of view, though perhaps not from another) may be inevitable in the sense that the guilty liberalism is *also* a precondition of progress.

Finally, let me repeat that for me the chauvinism of science is a much greater problem than the problem of intellectual pollution. It may even be one of its major causes. Scientists are not content with running their own playpens in accordance with what they regard as the rules of scientific method, they want to universalize these rules, they want them to become part of society at large and they use every means at their disposal – argument, propaganda, pressure tactics, intimidation, lobbying – to achieve their aims. The Chinese Communists recognized the dangers inherent in this chauvinism and they proceeded to remove it. In the process they restored important parts of the intellectual and emotional heritage of the Chinese people and they also improved the practice of medicine (cf. text to footnotes 9–13 of Chapter 4). It would be of advantage if other governments followed suit.

Appendix 4

Imre Lakatos has on various occasions reacted to the criticism contained in this chapter. He talked about it in lectures (such as his lectures at the Alpbach summer school of 1973), commented on it in letters and in private conversations. At one stage he seemed to say that while epistemological anarchism could not be killed by argument it could still be shown to be absurd: where is the epistemological anarchist who out of sheer contrariness walks out of the window of a 50-story building instead of using the lift? Towards the end of his life this seemed to be his main objection to me. I was puzzled by it for quite a while until I found what I think is a decisive answer. I wrote the answer on a piece of paper, nailed it to the wall next to my favourite chair and intended to use it as part of my reply to Imre's final criticism. The answer is as follows:

The case of the window-avoiding anarchist shows that anarchists often behave in a predictable way. It does not show that they, or their fellow window-avoiders, are guided by a rationality theory, for example, that they have chosen the behaviour suggested by the most advanced research programme known to them. Kittens approaching a painted abyss draw back, even if this is the first thing they see. Their behaviour, most likely, is innate. People draw back because they were trained to stay away from windows and because they firmly believe what to most of them can only be *rumours*, viz. reports on the deadly effects of high falls. Even the mechanical and physiological theories which the more wordy non-jumpers might use to justify their behaviour have not yet been shown to be in agreement with the methodology of research programmes and I doubt that it will ever be possible to remedy this situation. The epistemological anarchist, on the other hand, is not obliged to behave contrary to custom. He may readily admit that he is a coward, that he cannot control his fear, and that his fear keeps him away from windows.

(For details see Chapter 16, especially text to footnotes 38ff.) What he *does* deny is that he can give reasons for his fear which agree with the standards of some rationality theory, so that he is actually acting in accordance with standards. *This* is the point at issue and not what he does or does not do.

17

Moreover, these standards, which involve a comparison of content classes, are not always applicable. *The content classes of certain theories are incomparable in the sense that none of the usual logical relations (inclusion, exclusion, overlap) can be said to hold between them. This occurs when we compare myths with science. It also occurs in the most advanced, most general and therefore most mythological parts of science itself.*

I have much sympathy with the view, formulated clearly and elegantly by Whorff (and anticipated by Bacon), that languages and the reaction patterns they involve are not merely instruments for *describing* events (facts, states of affair), but that they are also *shapers* of events (facts, states of affair),[1] that their 'grammar' contains a cosmology, a comprehensive view of the world, of society, of the situation of man[2] which influences thought, behaviour, perception.[3] According to Whorff the cosmology of a language is expressed partly by the overt use of words, but it also rests on classifications 'which ha[ve] no overt mark . . . but which operate [ ] through an invisible "central exchange" of linkage bonds in such a way as to determine other words which mark the class.'[4] Thus '[t]he gender nouns, such as boy, girl, father, wife, uncle, woman, lady, including thousands of given names like George, Fred, Mary, Charlie, Isabel, Isadore, Jane, John, Alice, Aloysius, Esther, Lester,

1. According to Whorff 'the background linguistic system (in other words, the grammar) of each language is not merely a reproducing system for voicing ideas, but rather is itself a shaper of ideas, the programme and guide for the individual's mental activity, for his analysis of impressions, for his synthesis of his mental stock in trade'. *Language Thought and Reality*, MIT Press, 1956, p. 121. See also Appendix 5.

2. As an example cf. Whorff's analysis of Hopi Metaphysics in op. cit., pp. 57ff.

3. 'Users of markedly different grammars are pointed by their grammars towards different types of observations . . .', ibid., p. 221.

4. ibid., p. 69.

bear no distinguishing mark of gender like the Latin *-us* or *-a* within each motor process, but nevertheless each of these thousands of words has an invariable linkage bond connecting it with absolute precision either to the word "he" or to the word "she" which, however, does not come into the overt behaviour picture until and unless special situations of discourse require it.'[5]

Covert classifications (which, because of their subterranean nature are 'sensed rather than comprehended – awareness of [them] has an intuitive quality'[6] – which 'are quite apt to be more rational than overt ones'[7] and which may be very 'subtle' and not connected 'with any grand dichotomy'[8]) create 'patterned resistances to widely divergent points of view'.[9] If these resistances oppose not just the truth of the resisted alternatives but the presumption that an alternative has been presented, then we have an instance of incommensurability.

I also believe that scientific theories, such as Aristotle's theory of motion, the theory of relativity, the quantum theory, classical and modern cosmology are sufficiently general, sufficiently 'deep' and have developed in sufficiently complex ways to be considered along the same

5. ibid., p. 68.

6. ibid., p. 70. Even "[a] phoneme may assume definite semantic duties as part of its rapport. In English the phoneme ð ["thorn"] (the voiced sound of *th*) occurs initially only in the cryptotype [covert classification not connected with any grand dichotomy – p. 70] of demonstrative particles (the, this, there, than, etc.). Hence, there is a *psychic pressure* against accepting the voiced sound of *th* in new or imaginary words: *th*ig, *th*ay, *th*ob, *th*uzzle, etc., not having demonstrative meaning. Encountering such a new word (e.g. *th*ob) on a page, we will 'instinctively' give it the voiceless sound θ of *th* in 'think'. But it is not 'instinct'. Just our old friend linguistic rapport again" (p. 76, my italics).

7. ibid., p. 80. The passage continues: '. . . some rather formal and not very meaningful linguistic group, marked by some overt feature, may happen to coincide very roughly with some concatenation of phenomena in such a way as to suggest a rationalization of this parallelism. In the course of phonetic change, the distinguishing mark, ending, or what not is lost, and the class passes from a formal to a semantic one. Its reactance is now what distinguishes it as a class, and its idea is what unifies it. As time and use go on, it becomes increasingly organized around a rationale, it attracts semantically suitable words and loses former members that now are semantically inappropriate. Logic is now what holds it together.' Cf. also Mill's account of his educational development as described in text to footnote 14 of Chapter 12.

8. Whorff, op. cit., p. 70. Such subtle classifications are called *cryptotypes* by Whorff. A cryptotype is 'a submerged, subtle, and elusive meaning, corresponding to no actual word, yet shown by linguistic analysis to be functionally important in the grammar'.

9. ibid., p. 247.

lines as natural languages. The discussions that prepare the transition to a new age in physics, or in astronomy, are hardly ever restricted to the overt features of the orthodox point of view. They often reveal hidden ideas, replace them by ideas of a different kind, and change overt as well as covert classifications. Galileo's analysis of the tower argument led to a clearer formulation of the Aristotelian theory of space and it also revealed the difference between impetus (an absolute magnitude that inheres in the object) and momentum (which depends on the chosen reference system). Einstein's analysis of simultaneity unearthed some features of the Newtonian cosmology which, though unknown, had influenced all arguments about space and time, while Niels Bohr found in addition that the physical world could not be regarded as being entirely independent of the observer and gave content to the idea of independence that was part of classical physics.[10] Attending to cases such as these we realize that scientific arguments may indeed be subjected to 'patterned resistances'[11] and we expect that incommensurability will also occur among theories.

(As incommensurability depends on covert classifications and involves major conceptual changes it is hardly ever possible to give an explicit definition of it. Nor will the customary 'reconstructions' succeed in bringing it to the fore. The phenomenon must be shown, the reader must be led up to it by being confronted with a great variety of instances, and he must then judge for himself. This will be the method adopted in the present chapter.)[12]

Interesting cases of incommensurability occur already in the domain of *perception*. (This is not surprising if we remember the considerations in Chapter 14, above.) Given appropriate stimuli, but different systems

10. Cf. 'On a Recent Critique of Complementarity, Part II', *Philosophy of Science*, No. 36, 1969, pp. 92ff.

11. One recent example is Popper's criticism of Bohr. Popper is not troubled by any knowledge of Bohr's views (for proof cf. the article referred to in the previous note), and the position he attacks rests mostly in his own mind. But the *method* of attack shows to what extent he is still ruled by the ideology of classical physics (which plays a decisive role in his methodology as well as is seen from his definition of a basic statement in the *Logic of Scientific Discovery*, New York, 1959, p. 103: 'every basic statement must either be itself a statement about relative positions of physical bodies . . . or it must be equivalent to some basic statement of this "mechanistic" . . . kind'; i.e. basic statements are statements of classical physics).

12. Cf. footnote 6 and text.

of classification (different 'mental sets') our perceptual apparatus may produce perceptual objects which cannot be easily compared.[13] A direct judgement is impossible. We may compare the two attitudes in our *memory*, but *not* while attending to the *same picture*. The first drawing below goes one step further. It gives rise to perceptual objects which do

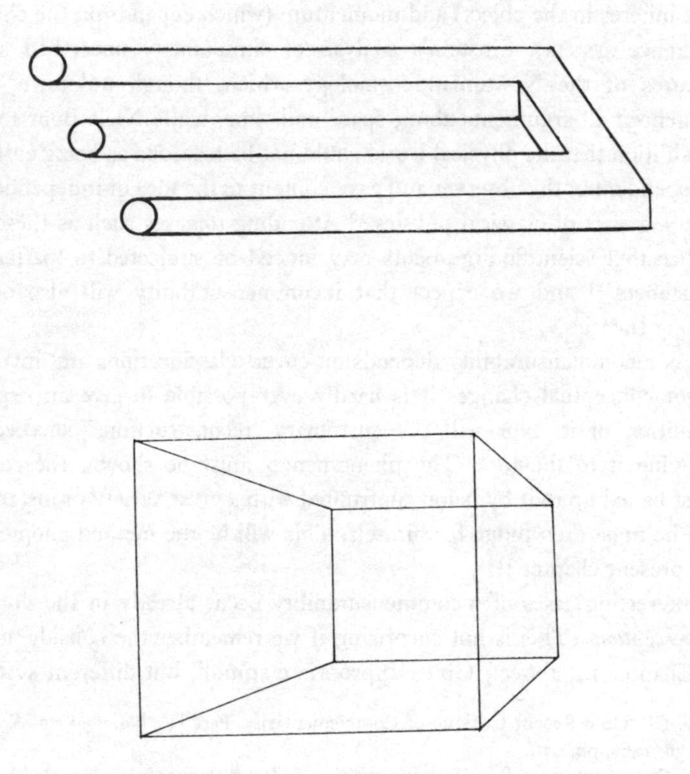

13. 'A master of introspection, Kenneth Clark, has recently described to us most vividly how even he was defeated when he attempted to "stall" an illusion. Looking at a great Velasquez, he wanted to observe what went on when the brush strokes and dabs of pigment on the canvas transformed themselves into a vision of transfigured reality as he stepped back. But try as he might, stepping backward and forward, he could never hold both visions at the same time . . .', E. Gombrich, *Art and Illusion*, Princeton, 1956, p. 6.

not just *negate* other perceptual objects – thus retaining the basic categories – but prevent the formation of any object whatsoever (note that the cylinder in the middle fades into nothingness as we approach the inside of the two-pronged stimulus). Not even memory can now give us a full view of the alternatives.

Every picture with only a modicum of perspective exhibits this phenomenon: we may decide to pay attention to the piece of paper on which the lines are drawn – but then there is no three-dimensional pattern; on the other hand we may decide to investigate the properties of this pattern, but then the surface of the paper disappears, or is integrated into what can only be called an illusion. There is no way of 'catching' the transition from the one to the other.[14] In all these cases the perceived image depends on 'mental sets' that can be changed at will, without the help of drugs, hypnosis, reconditioning. But mental sets may become frozen by illness, as a result of one's upbringing in a certain culture, or because of physiological determinants not in our control. (Not every change of language is accompanied by perceptual changes.) Our attitude towards other races, or towards people of a different cultural background often depends on 'frozen' sets of the second kind: having learned to 'read' faces in a standard way we make standard judgements and are led astray.

An interesting example of physiologically determined sets leading to incommensurability is provided by the *development of human perception*. As has been suggested by Piaget and his school,[15] a child's perception proceeds through various stages before it reaches its relatively stable adult form. In one stage, objects seem to behave very much like after-images and are treated as such. The child follows the object with his eyes until it disappears; he does not make the slightest attempt to recover it, even if this should require but a minimal physical (or intellectual) effort, an effort, moreover, that is already within the child's reach. There

14. Cf. R. L. Gregory, *The Intelligent Eye*, London, 1970, Chapter 2. Cf. also the distinction between eikon and phantasma in Plato, *Sophistes*, 235b8ff: 'This "appearing" or "seeming" without really "being" . . . all these expressions have always been and still are deeply involved in perplexity.' Plato talks about the distortions in statues of colossal size which were introduced to make them *appear* with the proper proportions. 'I cannot make use of an illusion and watch it,' says Gombrich in such cases, op. cit., p. 6.

15. J. Piaget, *The Construction of Reality in the Child*, New York, 1954, pp. 5ff.

is not even a tendency to search – and this is quite appropriate, 'conceptually' speaking. For it would indeed be nonsensical to 'look for' an after-image. Its 'concept' does not provide for such an operation.

The arrival of the concept, and of the perceptual image, of material objects, changes the situation quite dramatically. There occurs a drastic reorientation of behavioural patterns and, so one may conjecture, of thought. After-images, or things somewhat like them, still exist; but they are now difficult to find and must be discovered by special methods (the earlier visual world therefore literally *disappears*).[16] Such methods proceed from a new conceptual scheme (after-images occur in *humans*, they are not parts of the physical world) and cannot lead back to the exact phenomena of the previous stage. (These phenomena should therefore be called by a different name, such as 'pseudo-after-images' – a very interesting perceptual analogue to the transition from, say, Newtonian mechanics to special relativity.) Neither after-images nor pseudo-after-images have a special position in the new world. For example, they are not treated as *evidence* on which the new notion of a material object is supposed to rest. Nor can they be used to *explain* this notion: after-images *arise together with it*, they depend on it, and are absent from the minds of those who do not yet recognize material objects; and pseudo-after-images *disappear* as soon as such recognition takes place. The perceptual field never contains after-images together with pseudo-after-images. It is to be admitted that every stage possesses a kind of observational 'basis' to which special attention is paid and from which a multitude of suggestions are received. However, this basis (a) *changes* from stage to stage, and (b) it is *part* of the conceptual apparatus of a given stage, not its one and only source of interpretation as some empiricists would like to make us believe.

Considering developments such as these, we may suspect that the family of concepts centring upon 'material object' and the family of concepts centring upon 'pseudo-after-image' are incommensurable in precisely the sense that is at issue here; these families cannot be used

16. This seems to be a general feature of the acquisition of new perceptual worlds: 'The older representations for the most part have to be suppressed rather than reformed,' writes Stratton in his epoch-making essay 'Vision without Inversion of the Retinal Image', *The Psychological Review*, IV, 1897, p. 471.

simultaneously and neither logical nor perceptual connections can be established between them.

Now is it reasonable to expect that conceptual and perceptual changes of this kind occur in childhood only? Should we welcome the fact, if it is a fact, that an adult is stuck with a stable perceptual world and an accompanying stable conceptual system, which he can modify in many ways but whose general outlines have forever become immobilized? Or is it not more realistic to assume that fundamental changes, entailing incommensurability, are still possible and that they should be encouraged lest we remain forever excluded from what might be a higher stage of knowledge and consciousness? Besides, the question of the mobility of the adult stage is at any rate an empirical question that must be attacked by *research*, and cannot be settled by methodological *fiat*.[17] The attempt to break through the boundaries of a given conceptual system, and to escape the reach of 'Popperian spectacles',[18] is an essential part of such research (it also should be an essential part of any interesting life).

Such an attempt involves much more than a prolonged 'critical discussion'[19] as some relics of the enlightenment would have us believe. One must be able to *produce* and to *grasp* new perceptual and conceptual relations, including relations which are not immediately apparent (covert relations – see above) and *that* cannot be achieved by a critical discussion alone (cf. also above, Chapters 1 and 2). The orthodox accounts, of course, are restricted to (physical) theories (or, rather, to emaciated caricatures of them),[20] they neglect the covert relations that contribute to their meaning, disregard perceptual changes and treat the rest in a rigidly standardized way so that any debate of unusual ideas is at once stopped by a series of routine responses. But now this whole array of

17. As Lakatos attempts to do: 'Falsification', p. 179, footnote 1: 'Incommensurable theories are neither inconsistent with each other, nor comparable for content. But we can *make* them, by a dictionary, inconsistent and their content comparable.'

18. 'Falsification', p. 177. 'Popperian spectacles' have, of course, not been invented by Popper, but are a common spiritual property of 18th-century *Aufkläricht*. Herder was the first to see their limitations (and to bring the wrath of Kant upon himself, as a result).

19. Popper in *Criticism and the Growth of Knowledge*, p. 56.

20. This is true for philosophy of science though not for general epistemology where one is content with examining the linguistic habits of that bedraggled, but still surviving late Stone Age creature, *Homo Oxoniensis*.

responses is in doubt. Every concept that occurs in it is suspect, especially 'fundamental' concepts such as 'observation', 'test', and, of course, the concept 'theory' itself. And as regards the word 'truth' we can at this stage only say that it certainly has people in a tizzy, but has not achieved much else. The best way to proceed in such circumstances is to use examples which are outside the range of the routine responses. It is for this reason that I have decided to examine means of representation different from languages or theories and to develop my terminology in connection with them. More especially, I shall examine styles in painting and drawing. It will emerge that there are no 'neutral' objects which can be represented in any style, and which can be used as objective arbiters between radically different styles. The application to languages is obvious.

The 'archaic style' as defined by Emanuel Loewy in his work on ancient Greek art[21] has the following characteristics.

(1) The structure and the movement of the figures and of their parts are limited to a few typical schemes; (2) the individual forms are stylized, they tend to have a certain regularity and are 'executed with . . . precise abstraction'[22]; (3) the representation of a form depends on the *contour* which may retain the value of an independent line or form the boundaries of a silhouette. 'The silhouettes could be given a number of postures: they could stand, march, row, drive, fight, die, lament. . . . But always their essential structure must be clear';[23] (4) *colour* appears in one shade only, and gradations of light and shadow are missing; (5) as a rule the figures show their parts (and the larger episodes their

21. *Die Naturwiedergabe in der älteren Griechischen Kunst*, Roma, 1900, Chapter 1. Loewy uses 'archaic' as a *generic* term covering phenomena in Egyptian, Greek and Primitive Art, in the drawings of children and of untutored observers. In Greece his remarks apply to the *geometric style* (1000 to 700 B.C.) down to the *archaic period* (700 to 500 B.C.) which treats the human figure in greater detail and involves it in lively episodes. Cf. also F. Matz, *Geschichte der Griechischen Kunst*, Vol. I, 1950, as well as Beazly and Ashmole, *Greek Sculpture and Painting*, Cambridge, 1966, Chapters II and III.

22. Webster, *From Mycenae to Homer*, New York, 1964, p. 292. Webster regards this use of 'simple and clear patterns' in Greek geometric art as 'the forerunner of later developments in art (ultimately the invention of perspective), mathematics, and, philosophy'.

23. Webster, op. cit., p. 205.

elements) *in their most complete aspect* – even if this means awkwardness in composition, and 'a certain disregard of spatial relationships'. The parts are given their known value even when this conflicts with their seen relationship to the whole;[24] thus (6) with a few well-determined exceptions the figures which form a composition are arranged in such a way that *overlaps are avoided* and objects situated behind each other are presented as being side by side; (7) the *environment* of an action (mountains, clouds, trees, etc.) is either completely disregarded or it is omitted to a large extent. The action forms self-contained units of typical scenes (battles, funerals, etc.).[25]

These stylistic elements which are found, in various modifications, in the drawings of children, in the 'frontal' art of the Egyptians, in early Greek art, as well as among Primitives, are explained by Loewy on the basis of psychological mechanisms: 'Side by side with the images which reality presents to the physical eye there exists an entirely different world of images which live or, better, come to life in our mind only and which, although suggested by reality, are totally transformed. Every primitive act of drawing . . . tries to reproduce these images and them alone with the instinctive regularity of a psychical function.'[26] The archaic style changes as a result of 'numerous planned observations of nature which modify the pure mental images',[27] initiate the development towards realism and thus start the history of art. *Natural*, physiological reasons are given for the archaic style and for its change.

Now it is not clear why it should be more 'natural' to copy memory images than images of perception which are so much better defined and so much more permanent.[28] We also find that realism often *precedes* more schematic forms of presentation. This is true of the Old Stone Age,[29]

24. ibid., p. 207.

25. Beazly and Ashmole, op. cit., p. 3.

26. Loewy, op. cit., p. 4.

27. ibid., p. 6.

28. The facts of perspective are noticed, but they do not enter the pictorial presentation; this is seen from literary descriptions. Cf. H. Schäfer, *Von Aegyptischer Kunst*, Wiesbaden, 1963, pp. 88ff, where the problem is further discussed.

29. Cf. Paolo Graziosi, *Palaeolithic Art*, New York, 1960, and André Leroc-Gourhan, *Treasures of Prehistoric Art*, New York, 1967, both with excellent illustrations. These results were not known to Loewy: Cartailhac's 'Mea culpa d'un sceptique', for example appeared only in 1902.

of Egyptian Art,[30] of Attic Geometric Art.[31] In all these cases the 'archaic style' is the result of a *conscious effort* (which may of course be aided, or hindered, by unconscious tendencies and physiological laws) rather than a natural reaction to internal deposits of external stimuli.[32] Instead of looking for the psychological *causes* of a 'style' we should therefore rather try to discover its *elements*, analyse their *function*, compare them with other phenomena of the same culture (literary style, sentence construction, grammar, ideology) and thus arrive at an outline of the underlying *world view* including an account of the way in which this world view influences perception, thought, argument, and of the limits it imposes on the roaming about of the imagination. We shall see that such an analysis of outlines provides a better understanding of the process of conceptual change than either the naturalistic account, or trite phrases such as 'a critical discussion and a comparison of . . . various frameworks is always possible'.[33] Of course, *some* kind of comparison is *always* possible (for example, one physical theory may sound more melodious when read aloud to the accompaniment of a guitar than another physical theory). But lay down specific rules for the process of comparison, such as the rules of logic as applied to the relation of content classes, and you will find exceptions, undue restrictions, and you will be forced to talk your way out of trouble at every turn. It is much more interesting and instructive to examine what kinds of things can be said

30. Cf. the change in the presentation of animals in the course of the transition from predynastic times to the First Dynasty. The Berlin lion (Berlin, Staatliches Museum, Nr. 22440) is wild, threatening, quite different in expression and execution from the majestic animal of the Second and Third Dynasties. The latter seems to be more a representation of the *concept* lion than of any individual lion. Cf. also the difference between the falcon on the victory tablet of King Narmer (backside) and on the burial stone of King Wadji (Djet) of the First Dynasty. 'Everywhere one advanced to pure clarity, the forms were strengthened and made simple,' Schäfer, pp. 12ff, especially p. 15 where further details are given.

31. 'Attic geometric art should not be called primitive, although it has not the kind of photographic realism which literary scholars seem to demand in painting. It is a highly sophisticated art with its own conventions which serve its own purposes. As with the shapes and the ornamentation, a revolution separates it from late Mycenaean painting. In this revolution figures were reduced to their minimum silhouettes, and out of these minimum silhouettes the new art was built up.' Webster, op. cit., p. 205.

32. This thesis is further supported by the observation that so-called Primitives often turn their back to the objects they want to draw; Schäfer, p. 102, after Conze.

33. Popper, in *Criticism*, etc., p. 56.

(represented) and what kinds of things cannot be said (represented) *if the comparison has to take place within a certain specified and historically well-entrenched framework*. For such an examination we must go beyond generalities and study frameworks in detail. I start with an account of some examples of the archaic style.

Illustrations B and C (see the end of this chapter, pp. 289ff) show the following characteristics of the human figure: 'the men are very tall and thin, the trunk of a triangle tapering to the waist, the head a knob with a mere excrescence for a face: towards the end of the style the head is lit up – the head knob is drawn in outline, and a dot signifies the eye'.[34] All, or almost all, parts are shown in profile and they are strung together like the limbs of a puppet or a rag doll. They are not 'integrated' to form an organic whole. This 'additive' feature of the archaic style becomes very clear from the treatment of the eye. The eye does not participate in the actions of the body, it does not guide the body or establish contact with the surrounding situation, it does not 'look'. It is added on to the profile head like part of a notation as if the artist wanted to say: 'and beside all these other things such as legs, arms, feet, a man has also eyes, they are in the head, one on each side' (Illustrations D and A contain the 'frontal eye'). Similarly, special states of the body (alive, dead, sick) are not indicated by a special arrangement of its parts, but by putting the same standard body into various standard *positions*. Thus the body of the dead man in the funeral carriage of Illustration C is articulated in exactly the same way as a standing man, but it is rotated through 90 degrees and inserted in the space between the bottom of the shroud and the top of the bier.[35] Being shaped like the body of a live man it is *in addition* put into the death position. Another instance is the picture of a kid half swallowed by a lion.[36] The lion looks ferocious, the kid looks peaceful, and the act of swallowing is simply *tacked on* to the presentation of what a lion *is* and what a kid *is*. (We have what is called a *paratactic aggregate*: the elements of such an aggregate are all given equal importance, the only relation between them is sequential, there is

34. Beazly and Ashmole, op. cit., p. 3.

35. Webster, p. 204: 'The painter feels the need to say that he has two arms, two legs, and a manly chest.'

36. R. Hampl, *Die Gleichnisse Homers und die Bildkunst seiner Zeit*, Tübingen, 1952.

no hierarchy, no part is presented as being subordinate to and determined by others.) The picture *reads*: ferocious lion, peaceful kid, swallowing of kid by lion.

The need to show every essential part of a situation often leads to a separation of parts which are actually in contact. The picture becomes a map. Thus the charioteer in Illustration E is shown as standing above the floor (which is presented in its fullest view) and unencumbered by the rails so that his feet, the floor, the rails can all be clearly seen. No trouble arises if we regard the painting as a *visual catalogue* of the parts of an event rather than as an illusory rendering of the event itself (no trouble arises we *say*: his *feet* touched the *floor* which is *rectangular*, and he was surrounded by a *railing* . . .).[37] But such an interpretation must be *learned*, it cannot be simply read off the picture.

The amount of learning needed may be quite considerable. Some Egyptian drawings and paintings can be decoded only with the help of either the represented object itself or with the help of three-dimensional representations of it (statuary in the case of humans, animals, etc.). Using such information we learn that the chair in Figure A represents the object of Figure C and not the object of Figure B and that it must be read: 'chair with backrest and four legs, legs connected by support' where it is understood that only the front legs and the two back legs are so connected.[38] The interpretation of groups is complicated and some cases are not yet understood.[39]

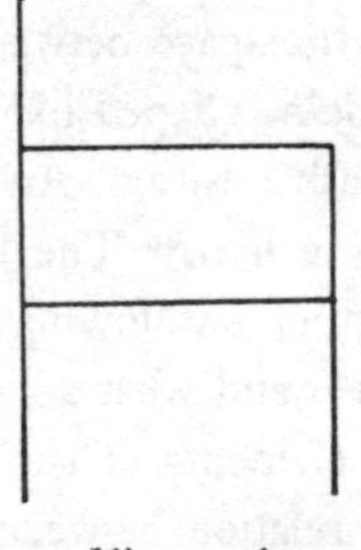

Figure A

37. 'All geometric pictures of chariots show at least one of these distortions.' Webster, p. 204. Late Mycenaean pottery, on the other hand, has the legs of the occupants concealed by the side.

38. Schäfer, op. cit., p. 123. 39. ibid., pp. 223ff.

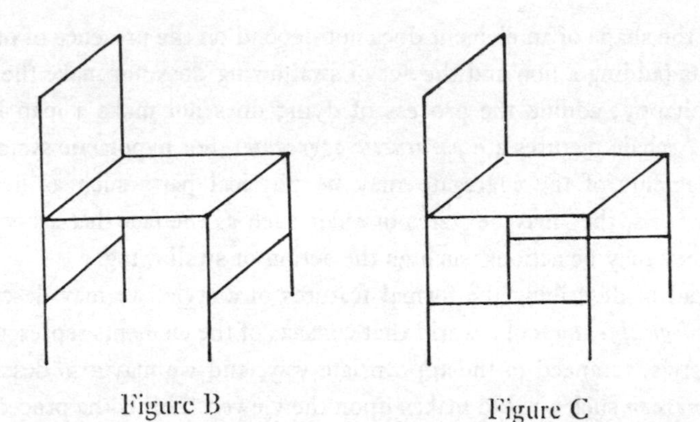

Figure B Figure C

(Being able to 'read' a certain style also includes knowledge of what features are *irrelevant*. Not every feature of an archaic list has representational value just as not every feature of a written sentence plays a role in articulating its content. This was overlooked by the Greeks who started enquiring into the reasons for the 'dignified postures' of Egyptian statues (already Plato commented on this). Such a question 'might have struck an Egyptian artist as it would strike us if somone enquired about the age or the mood of the king on the chessboard'.[40])

So far a brief account of some peculiarities of the 'archaic' style.

A style can be described and analysed in various ways. The descriptions given so far paid attention to *formal features*: the archaic style provides *visible lists* whose parts are arranged in roughly the same way in which they occur in 'nature' except when such an arrangement is liable to hide important elements. All parts are on the same level, we are supposed to 'read' the lists rather than 'see' them as illusory accounts of the situation.[41] The lists are not organized in any way except sequentially,

40. Gombrich, p. 134, with literature.

41. 'We come closer to the factual content of frontal [*geradvorstelliger*] drawings of objects, if we start by *reading off* their partial contents in the form of narrative declarative sentences. The frontal mode of representation gives us a "visual concept" [*Sehbegriff*] of the thing (the situation) represented.' Schäfer, op. cit., p. 118. Cf. also Webster, op. cit., p. 202, about the 'narrative' and 'explanatory' character of Mycenaean and geometric art. But cf. H. A. Groenewegen-Frankfort, *Arrest and Movement*, London, 1951, pp. 33f: the scenes from daily life on the walls of Egyptian tombs 'should be "read": harvesting entails ploughing, sowing, and reaping; care of cattle entails fording of streams and milking . . . the sequence of scenes is purely conceptual, not narrative,

that is, the shape of an element does not depend on the presence of other elements (adding a lion and the act of swallowing does not make the kid look unhappy; adding the process of dying does not make a man look weak). Archaic pictures are *paratactic aggregates*, not hypotactic systems. The elements of the aggregate may be physical parts such as heads, arms, wheels, they may be states of affair such as the fact that a body is dead, they may be actions, such as the action of swallowing.

Instead of describing the formal features of a style, we may describe the *ontological features* of a world that consists of the elements represented in the style, arranged in the appropriate way, and we may also describe the *impression* such a world makes upon the viewer. This is the procedure of the art critic who loves to dwell on the peculiar behaviour of the characters which the artist puts on his canvas and on the 'internal life' the behaviour seems to indicate. Thus G. M. S. Hanfmann[42] writes on the archaic figure: 'No matter how animated and agile archaic heroes may be, they do not appear to move by their own will. Their gestures are explanatory formulae imposed upon the actors from without in order to explain what sort of action is going on. And the crucial obstacle to the convincing portrayal of inner life was the curiously detached character of the archaic eye. It shows that a person is alive, but it cannot adjust itself to the demands of a specific situation. Even when the archaic artist succeeds in denoting a humorous or tragic mood, these factors of externalized gesture and detached glance recall the exaggerated animation of a puppet play.'

An ontological description frequently adds just verbiage to the formal analysis; it is nothing but an exercise in 'sensitivity' and cuteness. However, we must not disregard the possibility that a particular style *gives a precise account of the world as it is seen by the artist and his contemporaries* and that every formal feature corresponds to (hidden or explicit) assumptions inherent in the underlying cosmology. (In the

---

nor is the writing which occurs with the scenes dramatic in character. The signs, remarks, names, songs and explanations, which illuminate the action . . . do not link events or explain their development; they are typical sayings belonging to typical situations.'

42. 'Narration in Greek Art', *American Journal of Archaeology*, Vol. 61, January 1957, p. 74.

case of the 'archaic' style we must not disregard the possibility that man then actually *felt* himself to be a puppet guided by outside forces and that he *saw* and *treated* his fellow men accordingly.[43]) Such a *realistic interpretation* of styles and other means of representation would be in line with Whorff's thesis that in addition to being instruments for *describing* events (which may have other features, not covered by any description) languages are also *shapers* of events (so that there is a linguistic limit to what can be said in a given language, and this limit coincides with the limits of the thing itself).[44] The realistic interpretation is very plausible. But it must not be taken for granted.[45]

It must not be taken for granted, for there are technical failures, special purposes (caricature) which may change a style without changing the cosmology. We must also remember that all men have roughly the same neurophysiological equipment, so that perception cannot be bent in any direction one chooses.[46] And in some cases we can indeed show that deviations from a 'faithful rendering of nature' occur in the presence of a detailed knowledge of the object and side by side with more 'realistic' presentations: the workshop of the sculptor Thutmosis in Tel al-Amarna (the ancient Achet-Aton) contains masks directly taken from live models with all the details of the formation of the head (indentations) and of the face intact, as well as heads developed from such masks. Some of these heads preserve the details, others eliminate them and replace them by simple forms. An extreme example of such a style is the completely smooth head of an Egyptian man. It proves that 'at least some artists remained consciously independent of nature'.[47] During the reign of Amenophis IV (B.C. 1364–1347) the mode of representation was changed twice; the first change, towards a more realistic style, occurred merely

43. This is of course a very imprecise way of talking. One can have the 'impression of being a puppet' only if other impressions either occur, or are at least conceivable. Otherwise one just is what one is, without any specification.

44. Cf. footnote 1 and text of the present chapter.

45. For a sketch of the problems that arise in the case of *physical theories* cf. my 'Reply to Criticism', *Boston Studies in the Philosophy of Science*, Vol. 2, 1965, sections 5–8, and especially the list of problems on p. 234. Hanson, Popper and others take it for granted that realism is correct.

46. It may be different with drug-induced states, especially when they are made part of a systematic course of education. Cf. footnote 19 and text of the preceding chapter.

47. Schäfer, op. cit., p. 63.

four years after his ascension to the throne which shows that the technical ability for realism existed, was ready for use, but was intentionally left undeveloped. *An inference from style (or language) to cosmology and modes of perception therefore needs special argument; it cannot be made as a matter of course.* (A similar remark applies to any inference from popular theories in science, such as the theory of relativity, or the idea of the motion of the earth, to cosmology and modes of perception.)

The argument (which can never be conclusive) consists in pointing to characteristic features in distant fields. If the idiosyncrasies of a particular style of painting are found also in statuary, in the grammar of contemporary languages (and here especially in covert classifications which cannot be easily twisted around), if it can be shown that these languages are spoken by artists and by the common folk alike, if there are philosophical principles formulated in the languages which declare the idiosyncrasies to be features of the world and not just artifacts in it and which try to account for their origin, if man and nature have these features not only in paintings, but also in poetry, in popular sayings, in common law, if the idea that the features are parts of normal perception is not contradicted by anything we know from physiology, or from the psychology of perception, if later thinkers attack the idiosyncrasies as 'errors' resulting from an ignorance of the 'true way', then we may assume that we are not just dealing with technical failures and particular purposes, *but with a coherent way of life*, and we may expect that people involved in this way of life see the world in the same way in which we now see their pictures. It seems that all these conditions are satisfied in archaic Greece: the formal structure and the ideology of the *Greek epic* as reconstructed both from the text and from later references to it repeat all the peculiarities of the later geometric and the early archaic style.[48]

To start with, about nine-tenths of the Homeric epics consist of *formulae* which are prefabricated phrases extending in length from a single word or two to several complete lines and which are repeated at appropriate places.[49] One-fifth of the poems consist of lines wholly

48. Webster, op. cit., pp. 294ff.

49. In the 20th century the role of formulae was described and tested by Milman Parry, *L'Epithète traditionelle chez Homère*, Paris, 1928; *Harvard Studies in Classical Philology*, Vols. 41 (1930), 43 (1932). For a brief account cf. D. L. Page, *History and*

repeated from one place to another; in 28,000 Homeric lines there are about 25,000 repeated phrases. Repetitions occur already in Mycenaean court poetry and they can be traced to the poetry of eastern courts: 'Titles of gods, kings, and men must be given correctly, and in a courtly world the principle of correct expression may be extended further. Royal correspondence is highly formal, and this formality is extended beyond the messenger scenes of poetry to the formulae used for introducing speeches. Similarly, operations are reported in the terms of the operation order, whether the operation order itself is given or not, and this technique is extended to other descriptions, which have no such operation orders behind them. These compulsions all derive ultimately from the court of the king, and it is reasonable to suppose that the court in turn enjoyed such formality in poetry.'[50] The conditions of (Sumerian, Babylonian, Hurrian, Hethitic, Phoenician, Mycenaean) courts also explain the occurrence of standardized elements of *content* (typical scenes; the king and the nobles in war and peace; furniture; description of beautiful things) which, moving from city to city, and even across national boundaries are repeated, and adapted to local circumstances.

The slowly arising combination of constant and variable elements that is the result of numerous adaptations of this kind is utilized by the illiterate poets of the 'Dark Age' of Greece who develop a language and forms of expression that best serve the requirements of *oral composition.* The requirement of *memory* demands that there be ready-made descriptions of events that can be used by a poet who composes in his mind, and without the aid of writing. The requirement of *metre* demands that the basic descriptive phrases be fit for use in the various parts of the line the poet is about to complete: 'Unlike the poet who writes out his lines . . . [the oral poet] cannot think without hurry about his next word, nor change what he has made nor, before going on, read over what he has just written. . . . He must have for his use word groups all made to fit his verse.'[51] *Economy* demands that given a situation and a certain metrical constraint (beginning, middle or end of a line) there be

the Homeric Iliad, University of California Press, 1966, Chapter VI, as well as G. S. Kirk, *Homer and the Epic*, Cambridge, 1965, Part I.

50. Webster, op. cit., pp. 75f.

51. M. Parry, *Harvard Stud. Cl. Phil.*, 41, 1930, p. 77.

only one way of continuing the narration – and this demand is satisfied to a surprising extent: 'All the chief characters of the *Iliad* and the *Odyssey*, if their names can be fitted into the last half of the verse along with an epithet, have a noun-epithet formula in the nominative, beginning with a simple consonant, which fills the verse between the trochaic caesure of the third foot and the verse end: for instance, πολύτλας δῖος 'Οδυσσευς. In a list of thirty-seven characters who have formulae of this type, which includes all those having any importance in the poems, there are only three names which have a second formula which could replace the first.'[52] 'If you take in the five grammatic cases the singular of all the noun-epithet formulae used for Achilles, you will find that you have forty-five different formulae of which none has, in the same case, the same metrical value.'[53] Being provided for in this manner, the Homeric poet 'has no interest in originality of expression, or in variety. He uses or adapts inherited formulae.'[54] He does not have a 'choice, do[es] not even think in terms of choice; for a given part of the line, whatever declension case was needed, and whatever the subject matter might be, the formular vocabulary supplied at once a combination of words ready-made'.[55]

Using the formulae the Homeric poet gives an account of *typical scenes* in which objects are occasionally described by 'adding the parts on in a *string of words* in apposition'.[56] Ideas we would today regard as being logically subordinate to others are stated in separate, grammatically co-ordinate propositions. Example (*Iliad*, 9.556ff): Meleagros 'lay by his wedded wife, fair Cleopatra, daughter of fair-ankled Marpessa, daughter of Euenos, and of Ides, who was the strongest of men on earth at that time – and he against lord Phoebus Apollo took up his bow for the sake of the fair-ankled maid: her then in their halls did her father and lady mother call by the name of Alkyon because . . .' and so on, for ten more lines and two or three more major themes before a major stop. This *paratactic* feature of Homeric poetry which parallels the absence of elaborate systems of subordinate clauses in early Greek[57] also makes it

52. ibid., pp. 86f.
53. ibid., p. 89.
54. Page, op. cit., p. 230.
55. ibid., p. 242.
56. Webster, op. cit., pp. 99f; my italics.
57. Cf. Raphael Kühner, *Ausführliche Grammatik der Griechischen Sprache*, 2. Teil, reprinted Darmstadt, 1966. In the 20th century such a paratactic or 'simultanistic,

clear why Aphrodite is called 'sweetly laughing' when in fact she complains tearfully (*Iliad*, 5.375), or why Achilles is called 'swift footed 'when he is sitting talking to Priam (*Iliad*, 24.559). Just as in late geometric pottery (in the 'archaic' style of Loewy) a dead body is a live body brought into the position of death (cf. above, text to footnote 35) or an eaten kid a live and *peaceful kid* brought into the appropriate relation to the mouth of a ferocious lion, in the very same way Aphrodite complaining is simply Aphrodite – and that is the laughing goddess – *inserted* into the situation of complaining in which she participates only externally, without changing her nature.

The *additive treatment* of events becomes very clear in the case of (human) motion. In *Iliad*, 22.298, Achilles drags Hector along in the dust 'and dust arose around him that was dragged, and his dark hair flowed loose on either side, and in the dust *lay* his once fair head' – that is, the *process* of dragging contains the *state* of lying as an independent part which together with other such parts constitutes the motion.[58]

---

way of presentation was used by the early expressionists, for example by Jacob von Hoddis in his poem *Weltende*:

> Dem Bürger fliegt vom spitzen Kopf der Hut,
> In allen Lüften hallt es wie Geschrei.
> Dachdecker stürzen ab und gehn entzwei,
> Und an den Küsten – liest man – steigt die Flut.
>
> Der Sturm ist da, die wilden Meere hupfen
> An Land, um dicke Dämme zu zerdrücken.
> Die meisten Menschen haben einen Schnupfen.
> Die Eisenbahnen fallen von den Brücken.

Von Hoddis claims Homer as a precursor, explaining that simultaneity was used by Homer not in order to make an event more transparent but in order to create a feeling of immeasurable spaciousness. When Homer describes a battle and compares the noise of the weapons with the beat of a woodcutter, he merely wants to show that while there is battle there is also the quietness of woods, interrupted only by the work of the woodcutter. Catastrophe cannot be thought without simultaneously thinking of some utterly unimportant event. The Great is mixed up with the Small, the Important with the Trivial. (For the report cf. J. R. Becher in *Expressionismus*, ed. P. Raabe, Olten und Freiburg, 1965, pp. 50ff; this short article also contains a description of the tremendous impression von Hoddis' eight-liner made when it first came out in 1911.) One cannot infer that the same impression was created in the listener of the Homeric singers who did not possess a complex and romanticizing medium that had deteriorated into tearful sentimentality as a background for comparison.

58. Cf. Gebhard Kurz, *Darstellungsformen menschlicher Bewegung in der Ilias*, Heidelberg, 1966, p. 50.

Speaking more abstractly, we might say that for the poet 'time is composed of moments'.[59] Many of the similes assume that the parts of a complex entity have a life of their own and can be separated with ease. Geometrical man is a visible list of parts and positions, Homeric man is put together from limbs, surfaces, connections which are isolated by comparing them with inanimate objects of precisely defined shape: the trunk of Hippolochos rolls through the battle field like a *log* after Agamemnon has cut off his arms and his head (*Iliad*, 11.146 – ὅλμος, round stone of cylindrical shape), the body of Hector spins like a *top* (*Iliad*, 14.412), the head of Gorgythion drops to one side 'like a *garden poppy* being heavy with fruit and the showers of spring' (*Iliad*, 8.302)[60]; and so on. Also, the formulae of the epic, especially the noun-epithet combinations, are frequently used not according to content but according to metrical convenience: 'Zeus changes from counsellor to storm-mountain god to paternal god *not* in connection with what he is doing, but at the dictates of metre. He is not *nephelegerata Zeus* when he is gathering clouds, but when he is filling the metrical unit, ∪ ∪ — ∪ ∪ — —',[61] just as the geometrical artist may distort spatial relations – introduce contact where none exists and break it where it occurs – in order to tell the visual story in his own particular way. Thus the poet repeats the formal features used by the geometric and the early archaic artist. Neither seems to be aware of an 'underlying substance' that keeps the objects together and shapes their parts so that they reflect the 'higher unity' to which they belong.

Nor is such a 'higher unity' found in the *concepts* of the language. For example, there is no expression that could be used to describe the human body as a single entity.[62] *Soma* is the corpse, *demas* is accusative of specification, it means 'in structure', or 'as regards shape', reference to

59. This is the theory ascribed to Zeno by Aristotle, *Physics*, 239b, 31. The theory comes forth most clearly in the argument of the arrow: 'The arrow at flight is at rest. For, if everything is at rest when it occupies a space equal to itself, and what is in flight at any given moment always occupies a space equal to itself, it cannot move' (after *Physics*, 239b). We cannot say that the theory was held by Zeno himself, but we may conjecture that it played a role in Zeno's time.

60. Kurz, loc. cit.

61. R. Lattimore, *The Iliad of Homer*, Chicago, 1951, pp. 39f.

62. For the following cf. B. Snell, *The Discovery of the Mind*, Harper Torchbooks, 1960, Chapter 1.

limbs occurs where we today speak of the body (γυῖα, limbs as moved by the joints; μέλεα, limbs in their bodily strength; λέλυντο γυῖα, his whole body trembled; ἴδρος ἐκ μελέων ἔρρεν, his body was filled with strength). All we get is a puppet put together from more or less articulated parts.

This puppet does not have a soul in our sense. The 'body' is an aggregate of limbs, trunk, motion, the 'soul' is an aggregate of 'mental' events which are not necessarily private and which may belong to a different individual altogether. 'Never does Homer in his description of ideas or emotions, go beyond a purely spatial, or quantitative definition; never does he attempt to sound their special, non-physical nature.'[63] Actions are initiated not by an 'autonomous I', but by further actions, events, occurrences, including divine interference. And this is precisely how mental events are *experienced*.[64] Dreams, unusual psychological feats such as sudden remembering, sudden acts of recognition, sudden increase of vital energy, during battle, during a strenuous escape, sudden fits of anger are not only *explained* by reference to gods and demons, they are also *felt* as such. Agamemnon's dream 'listened to his [Zeus'] words and descended' (*Iliad*, 2.16) – the *dream* descends, not a figure in it 'and it stood then beside his [Agamemnon's] head in the likeness of Nestor' *Iliad*, 2.20). One does not *have* a dream (a dream is not a 'subjective' event), one *sees* it (it is an 'objective' event) and one also sees how it approaches and moves away.[65] Sudden anger, fits of strength are described *and felt* to be divine acts:[66] 'Zeus builds up and Zeus diminishes strength

63. Snell, op. cit., p. 18.

64. Cf. Dodds, *The Greeks and the Irrational*, Boston, 1957, Chapter I.

65. With some effort this experience can be repeated even today. Step 1: lie down, close your eyes, and attend to your hypnagogic hallucinations. Step 2: permit the hallucinations to proceed on their own and according to their own tendencies. They will then change from events in front of the eyes into events that gradually surround the viewer but without yet making him an active participant of an action in a three-dimensional dream-space. Step 3: switch over from *viewing* the hallucinatory event to *being part* of a complex of real events which act on the viewer and can be acted upon by him. Step 3 can be reversed either by the act of an almost non-existent will or by an outside noise. The three-dimensional scenery becomes two-dimensional, runs together into an area in front of the eyes, and moves away. It would be interesting to see how such *formal* elements change from culture to culture (so far one has examined dream-*content* only and formal elements only in so far as they are part of stage 3).

66. Today we say that somebody is 'overcome' by emotions and he may feel his

in man the way he pleases, since his power is beyond all others' (*Iliad*, 20.241) is not just an objective description (that may be extended to include the behaviour of animals) it also expresses the *feeling* that the change has entered from the outside, that one has been 'filled . . . with strong courage' (*Iliad*, 13.60). Today such events are either forgotten or regarded as purely accidental.[67] 'But for Homer, or for early thought in general, there is no such thing as accident.'[68] Every event is accounted for. This makes the events clearer, strengthens their objective features, moulds them into the shape of known gods and demons and thus turns them into powerful evidence for the divine apparatus that is used for explaining them: 'The gods are present. To recognize this as a given fact for the Greeks is the first condition for comprehending their religion and their culture. Our knowledge of their presence rests upon an (inner or outer) experience of either the Gods themselves or of an action of the Gods.'[69]

To sum up: the archaic world is much less compact than the world that surrounds us, and it is also experienced as being less compact. Archaic man lacks 'physical' unity, his 'body' consists of a multitude of parts, limbs, surfaces, connections; and he lacks 'mental' unity, his 'mind' is composed of a variety of events, some of them not even 'mental' in our sense, which either inhabit the body-puppet as additional constituents or are brought into it from the outside. Events are not *shaped* by the individual, they are complex arrangements of parts into which the body-puppet is *inserted* at the appropriate place.[70] This is the world-

anger as an alien thing that invades him against his will. The daemonic ontology of the Greeks contains objective terminology for describing this feature of our emotions *and thereby stabilizes it*.

67. Psychoanalysis and related ideologies now again contribute to making such events part of a wider context and thereby lend them substantiality.

68. Dodds, op. cit., p. 6.

69. Wilamowitz-Moellendorf, *Der Glaube der Hellenen*, I, 1955, p. 17. Our conceptions of the world subdivide an otherwise uniform material and create differences in perceived brightness where objective brightness has no gradient. The same process is responsible for the ordering of the rather chaotic impressions of our inner life, leading to an (inner) perception of divine interference, and it may even introduce daemons, gods, sprites into the domain of outer perceptions. At any rate – there is a sufficient number of daemonic experiences not to reject this conjecture out of hand.

70. This means that success is not the result of an effort on the part of the individual but the fortunate fitting together of circumstances. This is expressed in words such as

view that emerges as a result of an analysis of the *formal* features of 'archaic' art and Homeric poetry, taken in conjunction with an analysis of the *concepts* which the Homeric poet used for describing what he sees. Its main features are *experienced* by the individuals using the concepts. *These individuals live indeed in the same kind of world that is depicted by their artists.*

Further evidence for the conjecture can be obtained from an examination of 'meta-attitudes' such as general religious attitudes and 'theories' of (attitudes to) knowledge.

For the lack of compactness just described reappears in the field of ideology. There is a *tolerance* in religious matters which later generations found morally and theoretically unacceptable and which even today is regarded as a manifestation of frivolous and simple minds.[71] Archaic man is a religious eclectic, he does not object to foreign gods and myths, he adds them to the existing furniture of the world without any attempt at synthesis, or a removal of contradictions. There are no priests, there is no dogma, there are no categorical statements about the gods, man, the world.[72] (This tolerance can still be found with the Ionian philosophers of nature who develop their ideas side by side with myth without trying to eliminate the latter.) There is no religious 'morality'

πράττειν, which seem to designate *activities*. Yet in Homer they emphasize not so much the effect of the agent as the fact that the result comes about in the right way, that the process that brings it about does not encounter too many disturbances; it fits into the other processes that surround it (in the Attic dialect εὐπράττω still means 'I am doing well'). Similarly τέυχειν emphasizes not so much the personal achievement as the fact that things go well, that they fit into their surroundings. The same is true of the acquisition of knowledge. 'Odysseus has seen a lot and experienced much, moreover, he is the πολυμήχανος who can always help himself in new ways, and, finally, he is the man who listens to his goddess Athena. The part of knowledge that is based on seeing is not really the result of his own activity and research, it rather happened to him while he was driven around by external circumstances. He is very different from Solon who, as Herodotus tells us, was the first to travel for theoretical reasons, because he was interested in research. In Odysseus the knowledge of many things is strangely separated from his activity in the field of the ἐπίσασθαι: this activity is restricted to finding means for reaching a certain aim, in order to save his life and the life of his associates.' B. Snell, *Die alten Griechen und Wir*, Göttingen, 1962, p. 48. In this place also a more detailed analysis of pertinent terms.

71. Example: F. Schachermayer, *Die frühe Klassik der Griechen*, Stuttgart, 1966.

72. Cf. Wilamowitz-Moellendorf, op. cit.

in our sense, nor are the gods abstract embodiments of eternal principles.[73] This they became later, during the archaic age and as a result they 'lost [their] humanity. Hence Olympianism in its moralized form tended to become a religion of fear, a tendency which is reflected in the religious vocabulary. There is no word for 'god-fearing' in the *Iliad*.'[74] This is how life is dehumanized by what some people are pleased to call 'moral progress' or 'scientific progress'.

Similar remarks apply to the 'theory of knowledge' that is implicit in this early world view. The Muses in *Iliad*, 2.284ff, have knowledge because they are *close* to things – they do not have to rely on rumours – and because they know all the *many* things that are of interest to the writer, one after the other. 'Quantity, not intensity is Homer's standard of judgement' and of knowledge,[75] as becomes clear from such words as πολύφρων and πολύμητις, 'much pondering' and 'much thinking', as well as from later criticisms such as 'Learning of many things [πολυμαθίη] does not teach intelligence'.[76] An interest in, and a wish to understand, *many amazing things* (such as earthquakes, eclipses of the sun and the moon, the paradoxical rising and falling of the Nile), each of them explained in its own particular way and *without* the use of universal principles persists in the coastal descriptions of the 8th and 7th (and later) centuries (which simply *enumerate* the tribes, tribal habits, and coastal formations that are successively met during the journey), and even a thinker such as Thales is satisfied with making many interesting observations and providing many explanations without trying to tie them together in a system.[77] (The first thinker to construct a 'system' was Anaximander who followed Hesiod.) *Knowledge* so conceived is not obtained by trying to grasp an essence behind the reports of the senses,

73. M. P. Nilsson, *A History of Greek Religion*, Oxford, 1949, p. 152.

74. Dodds, p. 35.

75. Snell, *The Discovery of the Mind*, p. 18.

76. Heraclitus, after Diogenes Laertius, IX, 1.

77. The idea that Thales used a principle expressing an underlying unity of natural phenomena and that he identified this principle with water is first found in Aristotle, *Metaphysics*, 983b6–12 and 26ff. A closer look at this and other passages and consultation of Herodotus suggests that he still belongs to the group of those thinkers who deal with numerous extraordinary phenomena, make numerous observations without tying them together in a system. Cf. the vivid presentation in F. Krafft, *Geschichte der Naturwissenschaften*, I, Freiburg, 1971, Chapter 3.

but by (1) putting the observer in the right position relative to the object (process, aggregate), by inserting him into the appropriate place in the complex pattern that constitutes the world, and (2) by adding up the elements which are noted under these circumstances. It is the result of a complex survey carried out from a suitable vantage point. One may doubt a vague report, or a fifth-hand account, but it is not possible to doubt what one can clearly see with one's own eyes. The *object* depicted or described is the proper arrangements of the elements which may include foreshortenings and other perspectoid phenomena.[78] The fact that an oar looks broken in water lacks here the sceptical force it assumes in another ideology.[79] Just as Achilles sitting does not make us doubt that he is swift-footed – as a matter of fact, we would start doubting his swiftness if it turned out that he is in principle incapable of sitting – in the very same way the bent oar does not make us doubt that it is perfectly straight in air – as a matter of fact, we would start doubting its straightness if it did not look bent in water.[80] The bent oar is not an *aspect* that contradicts another *aspect* so that our inquiry into the nature of the oar is frustrated, it is a particular *part* (situation) of the real oar that is not only *compatible* with its straightness, but that demands it. We see: the objects of knowledge are as additive as the visible lists of the archaic artist and the situations described by the archaic poet.

Nor is there any uniform conception of knowledge.[81] A great variety

78. Perspectoid phenomena are sometimes treated as if they were special properties of the objects depicted. For example, a container of the Old Kingdom (Ancient Egypt) has an indentation on top, indicating perspective, but the indentation is presented as a feature of the object itself, Schäfer, op. cit., p. 266. Some Greek artists try to find situations where perspective does not need to be considered. Thus the peculiarity of the so-called red-figure style that arises in about 530 B.C. 'does not so much consist in the fact that foreshortenings are drawn, but in the new and highly varied ways to circumvent them', E. Pfuhl, *Malerei und Zeichnung der Griechen*, Vol. I, Munich, 1923, p. 378.

79. Cf. the discussion in Chapter 1 of A. J. Ayer's *Foundations of Empirical Knowledge*. The example is well known to the ancient sceptics.

80. This is also the way in which J. L. Austin takes care of the case. Cf. *Sense and Sensibilia*, New York, 1962. It is clear that problems such as the 'problem of the existence of theoretical entities' cannot arise under these circumstances either. All these problems are *created* by the new approach that superseded the additive ideology of archaic and pre-archaic times.

81. B. Snell, *Die Ausdrücke für den Begriff des Wissens in der vorplatonischen Philosophie*, Berlin, 1924. A short account is given in Snell, *Die alten Griechen und wir*,

of words is used for expressing what we today regard as different forms of knowledge, or as different ways of acquiring knowledge. σοφία[82] means expertise in a certain profession (carpenter, singer, general, physician, charioteer, wrestler) including the arts (where it praises the artist not as an outstanding creator but as a master of his craft); εἴδεναι, literally 'having seen', refers to knowledge gained from inspection; συνίημι, especially in the *Iliad*, though often translated as 'listening' or 'understanding', is stronger, it contains the idea of following and obeying, one absorbs something and acts in accordance with it (hearing may play an important role). And so on. Many of these expressions entail a receptive attitude on the part of the knower, he repeats in his actions the behaviour of the things around him, he follows them,[83] he acts as befits an entity that is inserted at the place he occupies.

To repeat and to conclude: the modes of representation used during the early archaic period in Greece are not just reflections of incompetence or of special artistic interests, they give a faithful account of what are felt, seen, thought to be fundamental features of the world of archaic man. This world is an open world. Its elements are not formed or held together by an 'underlying substance', they are not appearances from which this substance may be inferred with difficulty. They occasionally coalesce to form assemblages. The relation of a single element to the assemblage to which it belongs is like the relation of a part to an aggregate of parts and not like the relation of a part to an overpowering whole. The particular aggregate called 'man' is visited, and occasionally inhabited by 'mental events'. Such events may reside in him, they may also enter from the outside. Like every other object man is an exchange station of influences rather than a unique source of action, an 'I' (Descartes' 'cogito' has no point of attack in this world, and his argument cannot even start). There is a great similarity between this view and Mach's cosmology except that the elements of the archaic world are recognizable physical and mental shapes and events while the elements used by Mach

---

pp. 41ff. Cf. also von Fritz, *Philosophie und sprachlicher Ausdruck bei Demokrit, Plato, und Aristoteles*, Leipzig-Paris-London, 1938.

82. Only occurrence in Homer, *Iliad*, 15, 42, concerning the σοφία of a carpenter (an 'expert carpenter' translates Lattimore).

83. Cf. Snell, *Ausdrücke*, p. 50.

are more abstract, they are as yet unknown *aims* of research, not its *object*. In sum, the representational units of the archaic world view admit of a realistic interpretation, they express a coherent ontology, and Whorff's observations apply.

At this point I interrupt my argument in order to make some comments which connect the preceding observations with problems in the philosophy of science.

1. It may be objected that foreshortenings and other indications of perspective are such obvious features of our perceptual world that they cannot have been absent from the perceptual world of the Ancients. The archaic manner of presentation is therefore incomplete, and its realistic interpretation incorrect.

Reply: Foreshortenings are not an obvious feature of our perceptual world unless special attention is drawn to them (in an age of photography and film this is rather frequently the case). Unless we are professional photographers, film-makers, painters we perceive *things*, not *aspects*. Moving swiftly among complex objects we notice much less change than a perception of aspects would permit. Aspects, foreshortenings, if they enter our consciousness at all are usually suppressed just as after-images are suppressed when the appropriate stage of perceptual development is completed[84] and they are noticed in special situations only.[85] In ancient Greece such special situations arose in the theatre, for the first-row viewers of the impressive productions of Aeschylus and Agatharchos, and there is indeed a school that ascribes to the theatre a decisive influence on the development of perspective.[86] Besides, why should the perceptual world of the ancient Greeks coincide with ours? It needs much more argument than reference to a non-existent form of perception to consolidate the objection.

2. The reader should take notice of the method that has been used for establishing the peculiarities of the archaic cosmology. In *principle* the method is identical with the method of an anthropologist who examines the world-view of an association of tribes. The differences,

84. Cf. footnote 15ff and text of the present chapter.

85. Cf. footnote 16.

86. Cf. Part II of Hedwig Kenner, *Das Theater und der Realismus in der Griechischen Kunst*, Vienna, 1954, especially pp. 121f.

which are quite noticeable, are due to the scarcity of the evidence and to the particular circumstances of its origin (written sources; works of art; no personal contact). Let us take a closer look at the method that is used in both cases!

An anthropologist trying to discover the cosmology of his chosen tribe and the way in which it is mirrored in language, in the arts, in daily life (question of realism vs. instrumentalism), first learns the language and the basic social habits; he inquires how they are related to other activities, including such *prima facie* unimportant activities as milking cows and cooking meals;[87] he tries to identify *key ideas.*[88] His attention to minutiae is not the result of a misguided urge for completeness but of the realization that what looks insignificant to one way of thinking (and perceiving) may play a most important role in another. (The differences between the paper-and-pencil operations of a Lorentzian and those of an Einsteinian are often minute, if discernible at all; yet they express a major clash of ideologies.)

Having found the key ideas the anthropologist tries to *understand* them. This he does in the same way in which he originally gained an understanding of his own language, including the language of the special profession that provides him with an income. He *internalizes* the ideas so that their connections are firmly engraved in his memory and his reactions, and can be produced at will. 'The native society has to be in the anthropologist himself and not merely in his notebooks if he is to understand it.'[89] *This process must be kept free from external interference.* For example, the researcher must not try to get a better hold on the ideas of the tribe by likening them to ideas he already knows, or finds more comprehensible or more precise. On no account must he attempt a 'logical reconstruction'. Such a procedure would tie him to the known, or to what is preferred by certain groups, and would forever prevent him from grasping the unknown ideology he is examining.

Having completed his study, the anthropologist carries within himself both the native society and his own background, and he may now start comparing the two. The comparison decides whether the native way of thinking can be reproduced in European terms (provided there is a

87. Evans-Pritchard, *Social Anthropology*, Free Press, 1965, p. 80.
88. ibid., p. 80.
89. ibid., p. 82.

unique set of 'European terms'), or whether it has a 'logic' of its own, not found in any Western language. In the course of the comparison the anthropologist may rephrase certain native ideas in English. This does not mean that English *as spoken independently of the comparison* is commensurable with the native idiom. It means that languages can be *bent* in many directions and that understanding does not depend on any particular set of rules.

3. The examination of key ideas passes through various stages, none of which leads to a complete clarification. Here the researcher must exercise firm control over his urge for instant clarity and logical perfection. He must never try to make a concept clearer than is suggested by the material (except as a temporary aid for further research). It is this material and not his logical intuition that decides about the content of the concepts. To take an example. The Nuer, a Nilotic tribe which has been examined by Evans-Pritchard, have some interesting spatio-temporal concepts.[90] The researcher who is not too familiar with Nuer thought will find the concepts 'unclear and insufficiently precise'. To improve matters he might try explicating them, using the notions of special relativity. That might create clear concepts, but they would no longer be Nuer concepts. If, on the other hand, he wants to get concepts which are both clear and Nuer, then he must keep his key notions vague and incomplete *until the right information comes along*, i.e., until field study turns up the missing elements which, taken by themselves, are just as unclear as the elements he has already found.

Each item of information is a building block of understanding, which means that it is to be clarified by the discovery of further blocks from the language and ideology of the tribe rather than by premature definitions. Statements such as '. . . the Nuer . . . cannot speak of time as though it was something actual, which passes, can be waited for, can be saved, and so forth. I do not think that they ever experience the same feeling of fighting against time, or of having to co-ordinate activities with an abstract passage of time, because their points of reference are mainly the activities themselves, which are generally of a leisurely character . . .'[91] are either building blocks – in this case their own content is incomplete and not

90. Evans-Pritchard, *The Nuer*, Oxford, 1940, Part III; cf. also the brief account in *Social Anthropology*, pp. 102ff.

91. *The Nuer*, p. 103.

fully understood – or they are preliminary attempts to anticipate the arrangement of the totality of all blocks. They are then to be tested, and elucidated by the discovery of further blocks rather than by logical clarifications (a child learns the meaning of a word not by logical clarification but by realizing how it goes together with things and other words). Lack of clarity of any particular anthropological statement indicates the scarcity of the material rather than the vagueness of the logical intuitions of the anthropologist.

4. Exactly the same remarks apply to my attempt to explore incommensurability. Within the sciences incommensurability is closely connected with meaning. A study of incommensurability in the sciences will therefore produce statements that contain meaning-terms – but these terms will be only incompletely understood, just as the term 'time' is incompletely understood in the quotation of the preceding paragraph. And the remark that such statements should be made only *after* production of a clear theory of meaning[92] is as sensible as the remark that statements about Nuer time, which are the material that *leads to* an understanding of Nuer time, should be written down only after such an understanding has been achieved. My argument presupposes, of course, that the anthropological method is the correct method for studying the structure of science (and, for that matter, of any other form of life).

5. Logicians are liable to object. They point out that an examination of meanings and of the relation between terms is the task of *logic*, not of anthropology. Now by 'logic' one may mean at least two different things. 'Logic' may mean the study of, or results of the study of, the structures inherent in a certain type of discourse. And it may mean a particular logical system, or set of systems.

A study of the first kind belongs to anthropology. For in order to see, for example, whether $AB \vee A\bar{B} \equiv A$ is part of the 'logic of quantum theory' we shall have to study quantum theory. And as quantum theory is not a divine emanation but the work of men, we shall have to study it in the form in which the work of men is usually available, that is, we shall

92. Achinstein, *Minnesota Studies in the Philosophy of Science*, 4, Minneapolis, 1970, p. 224, says that 'Feyerabend owe(s) us a theory of meaning' and Hempel is prepared to accept incommensurability only *after* the notion of meaning involved in it has been made clear, op. cit., p. 156.

have to study historical records – textbooks, original papers, records of meetings and private conversations, letters, and the like. (In the case of quantum theory our position is improved by the fact that the tribe of quantum theoreticians has not yet died out. Thus we can supplement historical study with anthropological field work.)

It is to be admitted that these records do not, by themselves, produce a *unique* solution to our problems.[93] But who has ever assumed that they do? Historical records do not produce a unique solution for historical problems either, and yet nobody suggests that they be neglected. There is no doubt that the records are *necessary* for a logical study in the sense examined now. The question is how they should be *used*.

We want to discover the structure of the field of discourse, of which the records give an incomplete account. We want to learn about it without changing it in any way. In our example we are not interested in whether a *perfected* quantum mechanics of the future employs $AB \vee A\bar{B} \equiv A$ or whether an *invention* of our own, whether a little bit of 'reconstruction' which changes the theory so that it conforms to some preconceived principles of modern logic and readily provides the answer employs that principle. We want to know whether quantum theory *as actually practised by physicists* employs the principle. For it is the work of the physicists and not the work of the reconstructionists we want to examine. And this work may well be full of contradictions and lacunae. Its 'logic' (in the sense in which I am now using the term) may well be 'illogical' when judged from the point of view of a particular system of formal logic.

Now, putting our question in this way we realize that it may not admit of any answer. There may not exist a single theory, one 'quantum theory', that is used in the same way by all physicists. The difference between Bohr and, say, von Neumann suggests that this is more than a distant possibility. To test the possibility, i.e. to either eliminate it or to give it shape, we must examine concrete cases. Such an examination of concrete cases may then lead to the result that quantum theoreticians differ from each other as widely as do Catholics and the various types of

93. In what follows I shall refer to two papers by J. Giedymin in *British Journal for the Philosophy of Science* of August 1970, pp. 257ff and February 1971, pp. 39ff. Reference will be by page number only. Giedymin says that logical problems cannot be *uniquely* solved by an analysis of historical documents and, so one may surmise, of anthropological records: p. 257.

Protestants: they have the same book (though even that is doubtful – just compare Dirac with von Neumann), but they sure are doing different things with it.

The need for anthropological case studies in a field that initially seemed to be dominated by a single myth, always the same, always used in the same manner, indicates that our common knowledge of science may be severely defective. It may be entirely mistaken (some mistakes have been hinted at in the preceding chapters). In these circumstances, the only safe way is to confess ignorance, to abandon reconstructions, and to start studying science from scratch. We must approach it like an anthropologist approaches the mental contortions of the medicine-men of a newly discovered association of tribes. And we must be prepared for the discovery that these contortions *are* wildly illogical (when judged from the point of view of formal logic) and *have to be* wildly illogical in order to function as they do.

6. Only a few philosophers of science interpret 'logic' in this sense, however. Only few philosophers are prepared to concede that the basic structures that underlie some newly discovered idiom might differ radically from the basic structures of the more familiar systems of formal logic and absolutely nobody is ready to admit that this might be true of science as well. Most of the time the 'logic' (in the sense discussed so far) of a particular language, or of a theory, is immediately identified with the features of a particular logical system without considering the need for an enquiry concerning the adequacy of such an identification. Professor Giedymin, for example, means by 'logic' a favourite system of his which is fairly comprehensive, but by no means all-embracing. (For example, it does not contain, nor could it be used to formulate, Hegel's ideas. And there have been mathematicians who have doubted that it can be used for expressing informal mathematics.) A logical study of science as Giedymin and his fellow logicians understand it, is a study of sets of formulae of this system, of their structure, the properties of their ultimate constituents (intension, extension, etc.), of their consequences and of possible models. If this study does not repeat the features an anthropologist has found in, say, science then this either shows that science has some faults, or that the anthropologist does not know any logic. It does not make the slightest difference to the logician in this second sense

that his formulae *do not look* like scientific statements, that they *are not used* like scientific statements and that science could not be possibly run in the simple ways his brain is capable of understanding (and therefore regards as the only permissible ways). He either does not notice the discrepancy or he regards it as due to imperfections that are to be removed from a satisfactory account. Not once does it occur to him that the 'imperfections' might have an important *function*, and that scientific progress might be impossible once they are removed. For him science *is* axiomatics plus model theory plus correspondence rules plus observation language.

Such a procedure assumes (without noticing that there is an assumption involved) that an anthropological study which familiarizes us with the overt and the hidden classifications of science has been completed, and that it has decided in favour of the axiomatic (etc., etc.) approach. No such study has ever been carried out. And the bits and pieces of field work available today, mainly as the result of the work of Hanson, Kuhn, Lakatos and others, show that the logician's approach removes not just some inessential embroideries of science, but those very features which make scientific progress and thereby science possible.

7. The discussions of meaning I have alluded to are another illustration of the deficiencies of the logician's approach. For Giedymin, who has written two long notes on the matter, this term and its derivatives, such as the term 'incommensurability', are 'unclear and insufficiently precise'.[94] I agree. Giedymin wants to make the terms clearer, he wants to understand them better. Again agreement. He tries to obtain the clarity he feels is lacking by explication in terms of a particular kind of formal logic and of the double language model, restricting the discussion to 'intension' and 'extension' as explained in the chosen logic. It is here that the disagreement starts. For the question is not how 'meaning' and 'incommensurability' occur within a particular logical system. The question is what role they play in (actual, i.e. non-reconstructed) science. Clarification must come from a more detailed study of this role, and lacunae must be filled with the results of such study. And as the filling takes time the key terms will be 'unclear and insufficiently precise' for years and perhaps decades. (See also items 3 and 4 above.)

94. Cf. footnote 93 for reference and context.

8. Logicians and philosophers of science do not see the situation in this way. Being both unwilling and unable to carry out an informal discussion, they demand that the main terms of the discussion be 'clarified'. And to 'clarify' the terms of a discussion does not mean to study the *additional* and as yet unknown properties of the domain in question which one needs to make them fully understood, it means to fill them with *existing* notions from the entirely different domain of logic and common sense, preferably observational ideas, until they sound common themselves, and to take care that the process of filling obeys the accepted laws of logic. The discussion is permitted to proceed only *after* its initial steps have been modified in this manner. So the course of an investigation is deflected into the narrow channels of things already understood and the possibility of fundamental conceptual discovery (or of fundamental conceptual change) is considerably reduced. Fundamental conceptual change, on the other hand, presupposes new world views and new languages capable of expressing them. Now, building a new world view, and a corresponding new language, is a process that takes considerable time, in science as well as in meta-science. The terms of the new language become clear only when the process is fairly advanced, so that each single word is the centre of numerous lines connecting it with other words, sentences, bits of reasoning, gestures which sound absurd at first but which become perfectly reasonable once the connections are made. Arguments, theories, terms, points of view and debates can therefore be clarified in at least two different ways: (a) in the manner already described, which leads back to the familiar ideas and treats the new as a special case of things already understood, and (b) by incorporation into a language of the future, which means *that one must learn to argue with unexplained terms and to use sentences for which no clear rules of usage are as yet available*. Just as a child who starts using words without yet understanding them, who adds more and more uncomprehended linguistic fragments to his playful activity, discovers the sense-giving principle only *after* he has been active in this way for a long time – the activity being a necessary presupposition of the final blossoming forth of sense – in the very same way the inventor of a new world view (and the philosopher of science who tries to understand his procedure) must be able to talk nonsense until the amount of nonsense created by him and

his friends is big enough to give sense to all its parts. There is again no better account of this process than the description which John Stuart Mill has left us of the vicissitudes of his education. Referring to the explanations which his father gave him on logical matters, he writes: 'The explanations did not make the matter at all clear to me at the time; but they were not therefore useless; they remained as a nucleus for my observations and reflections to crystallise upon; the import of his general remarks being interpreted to me, by the particular instances which came under my notice *afterwards*.'[95] Building a new language (for understanding the world, or knowledge) is a process of exactly the same kind *except* that the initial 'nuclei' are not given, but must be invented. We see here how essential it is to learn talking in riddles, and how disastrous an effect the drive for instant clarity must have on our understanding. (In addition, such a drive betrays a rather narrow and barbaric mentality: 'To use words and phrases in an easy going way without scrutinizing them too curiously is not, in general, a mark of ill breeding; on the contrary, there is something low bred in being too precise. . . .'[96])

All these remarks are rather trivial and can be illustrated by obvious examples. Classical logic arrived on the scene only when there was sufficient argumentative material (in mathematics, rhetorics, politics) to serve as a starting point and as a testing ground. Arithmetic developed without any clear understanding of the concept of number; such understanding arose only when there existed a sufficient amount of arithmetical 'facts' to give it substance. In the same way a proper theory of meaning (and of incommensurability) can arise only after a sufficient number of 'facts' has been assembled to make such a theory more than an exercise in concept-pushing. This is the reason for the examples in the present section.

9. There is still another dogma to be considered before returning to the main narration. It is the dogma that all subjects, however assembled,

95. There is much more randomness in this process than a rationalist would ever permit, or suspect, or even notice. Cf. von Kleist, 'Über die allmähliche Verfertigung der Gedanken beim Reden' in *Meisterwerke Deutscher Literaturkritik*, ed. Hans Meyer, Stuttgart, 1962, pp. 741–7. Hegel had an inkling of the situation. Cf. K. Loewith and J. Riedel ed., *Hegel, Studienausgabe I*, Frankfurt, 1968, p. 54.

96 Plato, *Theaitetos*, 184c. Cf. also I. Düring, *Aristoteles*, Heidelberg, 1966, p. 379, criticizing Aristotle's demand for instant precision.

quite automatically obey the laws of logic, or ought to obey the laws of logic. If this is so, then anthropological field work would seem to be superfluous. 'What is true in logic is true in psychology. . . . in scientific method, and in the history of science,' writes Popper.[97]

This dogmatic assertion is neither clear nor is it (in one of its main interpretations) true. To start with, assume that the expressions 'psychology', 'history of science', 'anthropology' refer to certain domains of facts and regularities (of nature, of perception, of the human mind, of society). Then the assertion is not *clear* as there is not a single subject – LOGIC – that might reveal the logical structure of these domains. There is Hegel, there is Brouwer, there are formalists. They offer not just different interpretations of one and the same bulk of logical 'facts', but different 'facts' altogether. And the assertion is not *true* as there exist legitimate scientific statements which violate simple logical rules. For example, there are statements which play an important role in established scientific disciplines and which are observationally adequate only if they are self-contradictory: fixate a moving pattern that has just come to a standstill, and you will see it move in the opposite direction, but without changing its position. The only phenomenologically adequate description is 'it moves, in space, but it does not change place' – and this description is self-contradictory.[98] There are examples from geometry:[99] thus the enclosed figure (which need not appear in the same way to every person) is seen as an isosceles triangle whose base is not halved by the perpendicular. And there are examples with $a = b \;\&\; b = c \;\&\; a \geqslant c$ as the only phenomenologically adequate description.[100] Moreover, there is not a single science, or other form of life that is useful, progressive as well as

97. *Objective Knowledge*, Oxford, 1972, p. 6. Anticipated e.g. by Comte, *Course*, 52e Leçon.

98. It has been objected (Ayer, G. E. L. Owen) that we are dealing with appearances, not with actual events, and that the correct description is 'it appears to move . . .'. But the difficulty remains. For if we introduce the 'appear', we must put it at the beginning of the sentence, which will read 'it appears that it moves and does not change place'. And as appearances belong to the domain of phenomenological psychology we have made our point, viz. that this domain contains self-inconsistent elements.

99. E. Rubin, 'Visual Figures Apparently Incompatible with Geometry', *Acta Psychologica*, VII, 1950, pp. 365ff.

100. E. Tranekjaer-Rasmussen, 'Perspectoid Distances', *Acta Psychologica*, XI, 1955, p. 297.

in agreement with logical demands. Every science contains theories which are inconsistent both with facts and with other theories and which reveal contradictions when analysed in detail. Only a dogmatic belief in the principles of an allegedly uniform discipline 'Logic' will make us disregard this situation.[101] And the objection that logical principles and principles of, say, arithmetic differ from empirical principles by not being accessible to the method of conjecture and refutations (or, for that matter, any other 'empirical' method) has been defused by more recent research in this field.[102]

Secondly, let us assume that the expressions 'psychology', 'anthropology', 'history of science', 'physics' do not refer to facts and laws, but to certain *methods* of assembling facts including certain ways of connecting

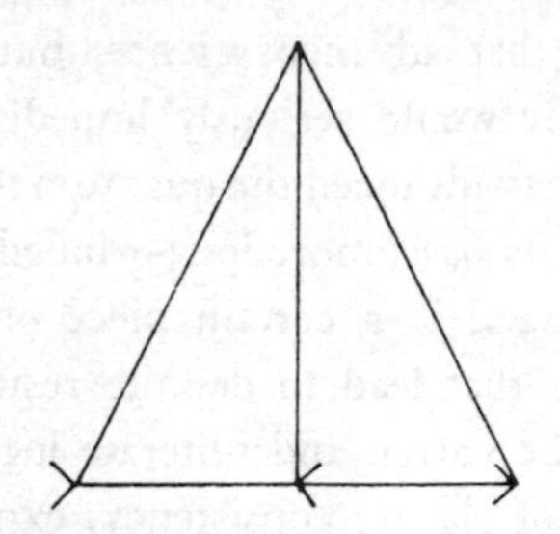

observation with theory and hypothesis. That is, let us consider the *activity* 'science' and its various subdivisions. We may approach this activity in two ways. We may lay down *ideal demands* of knowledge and knowledge-acquisition, and we may try to construct a (social) machinery that obeys these demands. Almost all epistemologists and philosophers of science proceed in this way. Occasionally they succeed in finding a machinery that might work in certain ideal conditions, but they never enquire, or even find it worth enquiring, whether the conditions are satisfied in this real world of ours. Such an enquiry, on the other hand, will have to explore the way in which scientists *actually* deal with their surroundings, it will have to examine the actual shape of their product, viz. 'knowledge', and the way in which this product changes as a result

101. Mach criticized the theory of relativity because it did not pay attention to psychological phenomena. Cf. the introduction to his *Physical Optics*.

102. Mainly by the work of Imre Lakatos, 'Proofs and Refutation', *British Journal for the Philosophy of Science*, 1962/63.

of decisions and actions in complex social and material conditions. In a word, such an enquiry will have to be anthropological.

There is no way of predicting what an anthropological enquiry will bring to light. In the preceding chapters, which are rough sketches of an anthropological study of particular episodes, it has emerged that science is always full of lacunae and contradictions, that ignorance, pigheadedness, reliance on prejudice, lying, far from impeding the forward march of knowledge are essential presuppositions of it and that the traditional virtues of precision, consistency, 'honesty', respect for facts, maximum knowledge under given circumstances, if practised with determination, may bring it to a standstill. It has also emerged that logical principles not only play a much smaller role in the (argumentative and non-argumentative) moves that advance science, but that the attempt to enforce them universally would seriously impede science. (One cannot say that von Neumann has advanced the quantum theory. But he certainly made the discussion of its basis more long-winded and cumbersome.)[103]

Now a scientist engaged in a certain piece of research has not yet completed all the steps that lead to definite results. His future is still open. Will he follow the barren and illiterate logician who preaches to him about the virtues of clarity, consistency, experimental support (or experimental falsification), tightness of argument, 'honesty', and so on, or will he imitate his predecessors in his own field who advanced by breaking most of the rules the logicians now want to lay on him? Will he rely on abstract injunctions or on the results of a study of concrete episodes? I think the answer is clear and with it the relevance of anthropological field work not just for the anthropologists but also for the members of the societies he examines.

I now continue my narration and proceed to describing the transition from the paratactic universe of the archaic Greeks to the substance-appearance universe of their followers.

The archaic cosmology (which from now on I shall call cosmology A) contains things, events, their parts; it does not contain any appearances.[104]

103. Besides, the imprecisions which he removes from the formalism now reappear in the relation between theory and fact. Here the correspondence principle still reigns supreme. Cf. footnote 23 of Chapter 5.

104. Snell, *Ausdrücke*, p. 28 (referring to Homer), speaks of a 'knowledge that pro-

Complete knowledge of an object is complete enumeration of its parts and peculiarities. Man cannot have complete knowledge. There are too many things, too many events, too many situations (*Iliad*, 2.488), and he can be close to only a few of them (*Iliad*, 2.485). But although man cannot have complete knowledge, he can have a sizeable amount of it. The wider his experience, the greater the number of adventures, of things seen, heard, read, the greater his knowledge.[105]

The new cosmology (cosmology B) that arises in the 7th to 5th centuries B.C. distinguishes between much-knowing, πολυμαθίη, and true knowledge,[106] and it warns against trusting 'custom born of manifold experience', ἔθος πολύπειρον.[107] Such a distinction and such a warning make sense only in a world whose structure is very different from the structure of A. In one version which played a large role in the development of Western civilization and which underlies such problems as the problem of the existence of theoretical entities and the problem of alienation the new events form what one might call a *True World*, while the events of everyday life are now *appearances* that are but its dim and misleading reflection.[108] The True World is simple and coherent, and it can be described in a uniform way. So can every act by which its elements are comprehended: a few abstract notions replace the numerous concepts that were used in cosmology A for describing how man might be 'inserted' into his surroundings and for expressing the equally numerous types of information thus gained. From now on there is only *one* important type of information, and that is: *knowledge*.

---

ceeds from appearances and draws their multitude together in a unit which is then posited as their true essence'. This may apply to the Presocratics, it does not apply to Homer. In the case of Homer 'the world is comprehended as the sum of things, visible in space, and not as reason acting intensively' (Snell, p. 67, discussing Empedokles; cf. also the lines following the quotation for a further elaboration of the theme).

105. Snell, *Die alten Griechen und Wir*, p. 48.

106. Cf. Heraclitus, fr. 40 (Diels-Kranz).

107. Parmenides, fr. 7, 3. 'Here for the first time sense and reason are contrasted'; W. K. Guthrie, *A History of Greek Philosophy*, Vol. II, Cambridge, 1965, p. 25.

108. This distinction is characteristic of certain mythological views as well. Homer is thus different both from the preceding mythologies and from the succeeding philosophies. His point of view is of great originality. In the 20th century J. L. Austin has developed similar ideas. And he has criticized the development from Thales via Plato to the present essentialism. Cf. the first chapter of *Sense and Sensibilia*.

The conceptual totalitarianism that arises as a result of the slow arrival of world B has interesting consequences, not all of them desirable. Situations which made sense when tied to a particular type of cognition now become isolated, unreasonable, apparently inconsistent with other situations: we have a 'chaos of appearances'. The 'chaos' is a direct consequence of the simplification of language that goes with the belief in a True World.[109] Moreover, all the manifold abilities of the observers are now directed towards this True World, they are adapted to a *uniform* aim, shaped for *one particular* purpose, they become more similar to each other which means that man becomes impoverished together with his language. He becomes impoverished at precisely the moment he discovers an autonomous 'I' and proceeds to what some have been pleased to call a 'more advanced notion of God' (allegedly found in Xenophanes), which is a notion of God lacking the rich variety of typically human features.[110] 'Mental' events which before were treated in analogy with events of the body and which *were experienced accordingly*[111] become more 'subjective', they become modifications, actions, revelations of a spontaneous soul: the distinction between appearance (first impression, mere opinion) and reality (true knowledge) spreads everywhere. Even the task of the *artist* now consists in arranging shapes in such a manner that the underlying essence can be grasped with ease. In painting this leads to the development of what one can only call systematic methods for deceiving the eye: the archaic artist treats the surface on which he paints as a writer might treat a piece of papyrus; it *is* a real surface, it is supposed to be *seen* as a real surface (though attention is not always directed to it) and the marks he draws on it are comparable to the lines of a blueprint or the letters of a word. They are symbols that *inform* the reader of the *structure of the object*, of its parts, of the way in which the parts are related to each other. The simple drawing opposite, for example, may represent three paths meeting at a point. The artist using perspective on the other hand, regards the surface and the marks he puts on it as *stimuli* that trigger the *illusion* of an arrangement of three-dimensional

109. Snell, *Ausdrücke*, pp. 80f; von Fritz, *Philosophie und sprachlicher Ausdruck bei Demokrit, Plato und Aristoteles*, Leipzig-Paris-London, 1938, p. 11.

110. '... in becoming the embodiment of cosmic justice Zeus lost his humanity. Hence Olympianism in its moralized form tended to become a religion of fear ...', Dodds, *Greeks*, p. 35.

111. Snell, *Discovery*, p. 69.

objects. The illusion occurs because the human mind is capable of producing illusory experiences when properly stimulated. The drawing is now seen either as the corner of a cube that extends towards the viewer, or as the corner of a cube that points away from him (and is seen from below), or else as a plane floating above the surface of the paper carrying a two-dimensional drawing of three paths meeting.

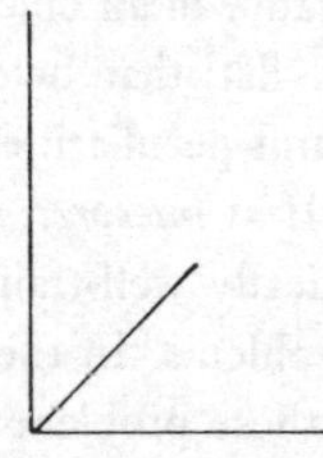

Combining this new way of seeing with the new concept of knowledge that has just been described, we obtain new entities, viz. physical objects as they are understood by most contemporary philosophers. To explain, let me again take the case of the oar.

In the archaic view 'the oar' is a complex consisting of parts some of which are objects, some situations, some events. It is possible to say 'the straight oar is broken' (*not* 'appears to be broken') just as it is possible to say 'swift-footed Achilles is walking slowly', for the elements are all of equal importance. They are part of a paratactic aggregate. Just as a traveller explores all parts of a strange country and describes them in a 'periegesis' that enumerates its peculiarities, one by one, in the same way the student of simple objects such as oars, boats, horses, people inserts himself into the 'major oar-situations', apprehends them in the appropriate way, and reports them in a list of properties, events, relations. And just as a detailed periegesis exhausts what can be said about a country, in the same way a detailed list exhausts what can be said about an object.[112] 'Broken in water' belongs to the oar as does 'straight to the

112. The idea that knowledge consists in *lists* reaches back far into the Sumerian past. Cf. von Soden, *Leistung und Grenzen Sumerisch-Babylonischer Wissenschaft*, Neuauflage, Darmstadt, 1965. The difference between Babylonian and Greek mathematics and astronomy lies precisely in this. The one develops methods for the presentation of what we today call 'phenomena' and which were interesting and relevant events in the sky, while the other tries to develop astronomy 'while leaving the heavens alone' (Plato, *Rep.*, 53aff; *Lgg.*, 818a).

hand'; it is 'equally real'. In cosmology B, however, 'broken in water' is a 'semblance' that *contradicts* what is suggested by the 'semblance' of straightness and thus shows the basic untrustworthiness of all semblances.[113] The concept of an object has changed from the concept of an aggregate of equi-important perceptible parts to the concept of an imperceptible essence underlying a multitude of deceptive phenomena. (We may guess that the appearance of an object has changed in a similar way, that objects now look less 'flat' than before.)

Considering these changes and peculiarities, it is plausible to assume that the comparison of A and B *as interpreted by the participants* (rather than as 'reconstructed' by logically well-trained but otherwise illiterate outsiders) will raise various problems. In the remainder of this chapter only some aspects of some of these problems will be discussed. Thus I shall barely mention the psychological changes that accompany the transition from A to B and which are not just a matter of conjecture,[114] but can be established by independent research. Here is rich material for the detailed study of the role of frameworks (mental sets, languages, modes of representation) and the limits of rationalism.

To start with, cosmos A and cosmos B are built from different *elements*.

The elements of A are relatively independent parts of objects which enter into external relations. They participate in aggregates without changing their intrinsic properties. The 'nature' of a particular aggregate is determined by its parts and by the way in which the parts are related to each other. *Enumerate the parts in the proper order, and you have the object*. This applies to physical aggregates, to humans (minds and bodies), to animals, but it also applies to social aggregates such as the honour of a warrior.

The elements of B fall into two classes: essences (objects) and appearances (of objects – what follows is true only of some rather streamlined versions of B). Objects (events, etc.) may again combine. They may form harmonious totalities where each part gives meaning to the whole and receives meaning from it (an extreme case is Parmenides where isolated parts are not only unrecognizable, but altogether unthinkable).

113. Xenophanes, fr. 34.
114. As similar changes are in most of Hanson's writings.

Aspects properly combined do not produce *objects*, but psychological conditions for the apprehension of *phantoms* which are but other aspects, and particularly misleading ones at that (they look so convincing). *No enumeration of aspects is identical with the object* (problem of induction).

The transition from A to B thus introduces new entities and new relations between entities (this is seen very clearly in painting and statuary). It also changes the concept and the self-experience of man. Archaic man is an assemblage of limbs, connections, trunk, neck, head,[115] he is a puppet set in motion by outside forces such as enemies, social circumstances, feelings (which are described and perceived as objective agencies – see above):[116] 'Man is an open target of a great many forces which impinge on him, and penetrate his very core.'[117] He is an exchange station of material and spiritual, but always objective, causes. And this is not just a 'theoretical' idea, it is a fact of observation. Man is not only *described* in this way, he is *pictured* in this way, and he *feels* himself to be constituted in this manner. He does not possess a central agency of action, a spontaneous 'I' that produces *its own* ideas, feelings, intentions, and differs from behaviour, social situations, 'mental' events of type A. Such an I is neither mentioned nor is it noticed. It is nowhere to be found within A. But it plays a very decisive role within B. Indeed, it is not implausible to assume that some outstanding peculiarities of B such as aspects, semblances, ambiguity of feeling[118] enter the stage as a result of a *sizeable increase of self-consciousness.*[119]

Now one might be inclined to explain the transition as follows: archaic man has a limited cosmology; he has discovered some things, he has missed others. His universe lacks important objects, his language lacks important concepts, his perception lacks important structures. Add the missing elements to cosmos A, the missing terms to language A, the

115. 'To be precise, Homer does not even have any words for the arms and the legs; he speaks of hands, lower arms, upper arms, feet, calves, and thighs. Nor is there a comprehensive term for the trunk.' Snell, *Discovery*, Chapter 1, footnote 7.

116. 'Emotions do not spring spontaneously from man, but are bestowed on him by the gods,' Snell, p. 52. See also the account earlier in the present chapter.

117. op. cit., p. 20.

118. Cf. Sappho's 'bitter-sweet Eros', Snell, p. 60.

119. For self-consciousness cf. Karl Pribram, 'Problems Concerning the Structure of Consciousness', MS, Stanford, 1973.

missing structures to the perceptual world of A, and you obtain cosmos B, language B, perception B.

Some time ago I called the theory underlying such an explanation the 'hole theory' or the 'Swiss cheese theory' of language (and other means of representation). According to the hole theory every cosmology (every language, every mode of perception) has sizeable lacunae which can be filled, *leaving everything else unchanged*. The hole theory is beset by numerous difficulties. In the present case there is the difficulty that cosmos B does not contain a single element of cosmos A. Neither common-sense terms, nor philosophical theories; neither painting and statuary, nor artistic conceptions; neither religion, nor theological speculation contain a single element of A once the transition to B has been completed. *This is a historical fact.*[120] Is this fact an accident, or has A some structural properties that prevent the co-existence of A-situations and B-situations? Let us see!

I have already mentioned an example that might give us an inkling of a reason as to why B does not have room for A-facts: the drawing below

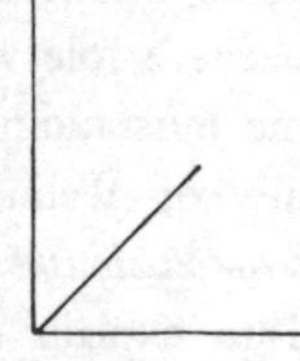

may be the intersection of three paths as presented in accordance with the principles of A-pictures (which are visual lists). Perspective having been introduced (either as an objective method or as a mental set), it can no longer be seen in this manner. Instead of lines on paper we have the illusion of depth and a three-dimensional panorama, though of a rather simple kind. There is no way of incorporating the A-picture into the B-picture except as part of this illusion. But an illusion of a visual list is not a visual list.

The situation becomes more transparent when we turn to concepts.

120. The fact is not easy to establish. Many presentations of A, including some very detailed and sophisticated ones, are infected by B-concepts. An example is quoted in footnote 104 to the present chapter. Here as elsewhere only the anthropological method can lead to knowledge that is more than a reflection of wishful thinking.

I have said above that the 'nature' of an object (= aggregate) in A is determined by the elements of the aggregate and the relation between the elements. One should add that this determination is 'closed' in the sense that the elements and their relations *constitute* the object; when they are given, then the object is given as well. For example, the 'elements' described by Odysseus in his speech in *Iliad*, 9.225ff *constitute* honour, grace, respect. A-concepts are thus very similar to notions such as 'checkmate': given a certain arrangement of pieces on the board, there is no way of 'discovering' that the game can still be continued. Such a 'discovery' would not fill a gap, it would not add to our knowledge of possible chess positions, it would put an end to the game. And so would the 'discovery' of 'real meanings' behind other moves and other constellations.

Precisely the same remarks apply to the 'discovery' of an individual I that is different from faces, behaviour, objective 'mental states' of the type that occur in A, to the 'discovery' of a substance behind 'appearances' (formerly elements of A), or to the 'discovery' that honour may be lacking despite the presence of all its outer manifestations. A statement such as Heraclitus' 'you could not find the limits of the soul though you are travelling every way, so deep is its *logos*' (Diels, p. 45) does not just *add* to cosmos A, it *undercuts* the principles which are needed in the construction of A-type 'mental states' while Heraclitus' rejection of πολυμαθίη and Parmenides' rejection of an ἔθος πολύπειρον undercuts rules that govern the construction of *every single fact* of A. An entire world view, an entire universe of thought, speech, perception is dissolved.

It is interesting to see how this process of dissolving manifests itself in particular cases. In his long speech in *Iliad*, 9.308ff, Achilles wants to say that honour may be absent even though all its outer manifestations are present. The terms of the language he uses are so intimately tied to definite social situations that he 'has no language to express his disillusionment. Yet he expresses it, and in a remarkable way. He does it by misusing the language he disposes of. He asks questions that cannot be answered and makes demands that cannot be met.'[121] He acts in a most 'irrational' way.

The same irrationality is found in the writings of all other early

121. A. Parry, 'The Language of Achilles', *Trans. & Proc. Amer. Phil. Assoc.*, 87, 1956, p. 6.

authors. Compared with A the Presocratics speak strangely indeed. So do the lyrical poets who explore the new possibilities of selfhood they have 'discovered'. Freed from the fetters of a well-constructed and unambiguous mode of expression and thinking, the elements of A lose their familiar function and start floating around aimlessly – the 'chaos of sensations' arises. Freed from firm and unambiguous social situations feelings become fleeting, ambivalent, contradictory: 'I love, and I love not; I rave, and I do not rave,' writes Anakreon.[122] Freed from the rules of late geometric painting the artists produce strange mixtures of perspective and blueprint.[123] Separated from well-determined psychological sets and freed of their realistic import, concepts may now be used 'hypothetically' without any odium of lying and the arts may begin exploring possible worlds in an imaginative way.[124] This is the same 'step back' which was earlier seen to be a necessary presupposition of change and, possibly, progress[125] – only it now does not just discard observations, it discards some important standards of rationality as well. Seen from A (and also from the point of view of some later ideologies) all these thinkers, poets, artists, are raving maniacs.

Remember the circumstances which are responsible for this situation.

122. Diehl, *Anthologia Lyrica*², fr. 79.

123. Pfuhl, op. cit.; cf. also J. White, *Perspective in Ancient Drawing and Painting*, London, 1965.

124. Plutarch reports the following story in his *Life of Solon*: 'When the company of Thespis began to exhibit tragedy, and its novelty was attracting the populace but had not yet got as far as public competitions, Solon, being fond of listening and learning and being rather given in his old age to leisure and amusement, and indeed to drinking parties and music, went to see Thespis act in his own play, as was the practice in ancient times. Solon approached him after the performance and asked him if he was not ashamed to tell so many lies to so many people. When Thespis said there was nothing dreadful in representing such works and actions in fun, Solon struck the ground violently with his walking stick: "If we applaud these things in fun," he said, "we shall soon find ourselves honouring them in earnest".' The story seems historically impossible yet elucidates a widespread attitude (for this attitude cf. Chapter 8 of John Forsdyke, *Greece before Homer*, New York, 1964). Solon himself seems to have been somewhat less impressed by traditional forms of thought and he may have been one of the first dramatic actors (of the political variety): cf. G. Else, *The Origin and Early Form of Tragedy*, Cambridge, 1965, pp. 40ff. The opposite attitude, which reveals the secure and already somewhat conceited citizen of B, is expressed by Simonides who answers the question why the Thessalians were not deceived by him by saying 'Because they are too stupid'. Plutarch, *De aud. poet.*, 15D.

125. Chapter 12, text to footnote 4.

We have a point of view (theory, framework, cosmos, mode of representation) whose elements (concepts, 'facts', pictures) are built up in accordance with certain principles of construction. The principles involve something like a 'closure': there are things that cannot be said, or 'discovered', without violating the principles (which does *not* mean contradicting them). Say the things, make the discovery, and the principles are suspended. Now take those constructive principles that underlie every element of the cosmos (of the theory), every fact (every concept). Let us call such principles *universal principles* of the theory in question. Suspending universal principles means suspending all facts and all concepts. Finally, let us call a discovery, or a statement, or an attitude *incommensurable* with the cosmos (the theory, the framework) if it suspends some of its universal principles. Heraclitus 45 is incommensurable with the psychological part of A: it suspends the rules that are needed for constituting individuals and puts an end to all A-facts about individuals (phenomena corresponding to such facts may of course persist for a considerable time as not all conceptual changes lead to changes in perception and as there exist conceptual changes that never leave a trace in the appearances;[126] however, such phenomena can no longer be *described* in the customary way and cannot therefore count as observations of the customary 'objective facts').

Note the tentative and vague nature of this explanation of 'incommensurable' and the absence of logical terminology. The reason for the vagueness has already been explained (items 3 and 4 above). The absence of logic is due to the fact that we deal with phenomena outside of its domain. My purpose is to find terminology for describing certain complex historical-anthropological phenomena which are only imperfectly understood rather than defining properties of logical systems that are specified in detail. Terms, such as 'universal principles' and 'suspend', are supposed to summarize anthropological information much in the same way in which Evans-Pritchard's account of Nuer time (text to footnote 91) summarizes the anthropological information at his disposal (cf. also the brief discussion in item 3 above). The vagueness of the explanation

126. This is overlooked by Hanson, who seems to expect that every major conceptual change will at once bend our perceptions. For details cf. footnote 52 and text of my 'Reply to Criticism', *Boston Studies in the Philosophy of Science*, Vol. II, New York, 1965.

reflects the incompleteness and complexity of the material and invites articulation by further research. The explanation has to have *some* content – otherwise it would be useless. But it must not have *too much* content, or else we have to revise it every second line.

Note, also, that by a 'principle' I do not simply mean a *statement* such as 'concepts apply when a finite number of conditions is satisfied', or 'knowledge is enumeration of discrete elements which form paratactic aggregates' but the *grammatical habit* corresponding to the statement. The two statements just quoted describe the habit of regarding an object as given when the list of its parts has been fully presented. This habit is suspended (though not contradicted) by the *conjecture* that even the most complete list does not exhaust an object; it is *also* suspended (but again not contradicted) by any unceasing search for new aspects and new properties. (It is therefore not feasible to define 'incommensurability' by reference to statements.)[127] If the habit is suspended, then A-objects are suspended with it: one cannot examine A-objects by a method of conjectures and refutations that knows no end.

How is the 'irrationality' of the transition period overcome? It is overcome in the usual way (cf. item 8 above), i.e. by the determined production of nonsense until the material produced is rich enough to permit the rebels to reveal, and everyone else to recognize, new universal principles. (Such revealing need not consist in writing the principles down in the form of clear and precise statements.) Madness turns into sanity provided it is sufficiently rich and sufficiently regular to function as the basis of a new world view. And when *that* happens, then we have a new problem: how can the old view be compared with the new view?

From what has been said it is obvious that we cannot compare the *contents* of A and B. A-facts and B-facts cannot be put side by side, not even in memory: presenting B-facts means suspending principles assumed in the construction of A-facts. All we can do is draw B-pictures of A-facts in B, or introduce B-statements of A-facts into B. We cannot use A-statements of A-facts in B. Nor is it possible to *translate* language A into language B. This does not mean that we cannot *discuss* the two

127. This takes care of a criticism in footnote 63 of Shapere's article in *Mind and Cosmos*, Pittsburgh, 1966. The classifications achieved by the principles are 'covert' in the sense of Whorff: cf. above, footnote 4 and text down to footnote 9.

views – but the discussion cannot be in terms of any (formal) logical relations between the elements of A and the elements of B. It will have to be as 'irrational' as was the talk of those intent on leaving A.

Now it seems to me that the relation between, say, classical mechanics (interpreted realistically) and quantum mechanics (interpreted in accordance with the views of Niels Bohr), or between Newtonian mechanics (interpreted realistically) and the general theory of relativity (also interpreted realistically) is in many respects similar to the relation between cosmology A and cosmology B. (There are of course also important differences; for example, the modern transition has left the arts, ordinary language, and perception unchanged.) Thus every fact of Newton's mechanics presumes that shapes, masses, periods are changed only by physical interactions and this presumption is suspended by the theory of relativity. Similarly the quantum theory constitutes facts in accordance with the uncertainty relations which are suspended by the classical approach.

I shall conclude this chapter by repeating its results in the form of theses. The theses may be regarded as summaries of anthropological material relevant for the elucidation, in accordance with items 3 and 4 above, of meaning-terms and of the notion of incommensurability.

The *first thesis* is that *there are* frameworks of thought (action, perception) which are incommensurable.

I repeat that this is a historical (anthropological) thesis which must be supported by historical (anthropological) evidence. For details cf. items 2 to 7 above. An example is provided by framework A and framework B.

It is, of course, always possible to replace a framework that looks strange and incomprehensible when approached from the point of view of western science by another framework that looks like some piece of western common sense (with, or without, science), or like a clumsy anticipation of such common sense, or else like a fantastic fairy-tale. Most early anthropologists distorted the object of their study in this way and so could easily assume that the English language (or the German, Latin or Greek language) was rich enough to present and comprehend the most outlandish myth. Early dictionaries express this belief in a very direct manner, for we have here simple definitions of all 'primitive' terms and simple explanations of all 'primitive' notions. In the meantime it has become clear that dictionaries and translations are lousy ways of introducing the concepts of a language that is not closely related to our own, or of ideas which do not fit into western ways of thinking.[128] Such languages have to be learned *from scratch*, as a child learns words, concepts, appearances[129] ('appearances' because things and faces are not just 'given', they are 'read' in certain ways – different ways being prominent in different ideologies). We must not demand that the process of learning be structured in accordance with the categories, laws and perceptions we are already familiar with. It is precisely such an 'unprejudiced' way of learning that a *field study* is supposed to achieve. Returning from the field study to his own concep-

128. Lakatos' observation (Falsification', op cit., p. 179, footnote 1) that 'we can *make*' incomparable views comparable by using 'a dictionary' still reflects the attitude of the older anthropologists. So does Giedymin's remark that 'any two languages and any two theories may be transformed into logically comparable ones' (*British Journal for the Philosophy of Science*, Vol. 21, 1970, p. 46), except that he adds the *proviso* 'if no restrictions on the extension of vocabulary and meaning rules are imposed'. In the case of anthropology we have, of course, an important restriction, and it is: keep as close as possible to the language that is being spoken by a certain tribe. In the philosophy of science the situation is exactly the same. We want to discover principles of scientific change. That is, we want to find out how Newton's theory *in the form in which it was available at about 1900* (when Einstein was looking for general physical principles that might be retained amidst the upheaval of classical ideas), is related to relativity *as conceived by* Einstein and not how *changed* versions of both Newton and Einstein (that can be expressed in the same language and, therefore, smoothly merge into each other) are related. I agree with Giedymin that 'rationalism as characterized by Popper requires that there be a common language to formulate critical argument' (p. 47). But my question is whether science – and that is the succession of fantastic theories as conceived by their inventors and not the bland reflection of this process in the minds of logicians and 'rationalists' – *knows* such a common language, and whether the attempt to *use* such a language would not bring it to a standstill. To answer *this* question we must look at science as it is and not as it looks after it has been made 'rational'. The answers we get in this way will most likely be unusual and exciting: scientists such as Einstein are unusual and exciting people, much more exciting than their logical 'explicators' can ever aspire to be.

129. For an interesting discussion of the situation in social anthropology cf. Chapter 4 of Part I of E. E. Evans-Pritchard, *Social Anthropology and Other Essays*, Free Press, 1964, especially pp. 82 top, 83 end of second paragraph, 85: 'people who belong to different cultures would notice different facts and perceive them in a different way. In so far as this is true, the facts recorded in our notebooks are not social facts, but ethnographic facts, selection and interpretation having taken place at the level of observation. . . .' This, of course, is also true of case studies of science, including those that use formidable logical machinery.

tions and his own language, such as English, an anthropologist often realizes that a direct translation has become impossible and that his views, and the views of the culture to which he belongs, are incommensurable with the 'primitive' ideas he has just begun to understand (or there may be overlap in some parts and incommensurability in others). Naturally, he wants to give an account of these ideas in English, but he will be able to do so only if he is prepared to use familiar terms in strange and novel ways. He may even have to build an entirely new language-game out of English words, and he will be able to start his explanations only when this language-game has become fairly complex. Now we know that almost every language contains within itself the means of restructuring large parts of its conceptual apparatus. Without this, popular science, science fiction, fairy-tales, tales of the supernatural, and science itself would be impossible. There is, therefore, a good sense in which we can say that the results of a field study can always be expressed in English. But this does not mean, as some self-styled rationalists seem to believe, that my first thesis is false. Such an inference could be justified only if it could be shown that a correct presentation (and not just a dictionary-caricature) of new views in a selected idiom, such as English, leaves the 'grammar' of this idiom unchanged. No such proof has ever been given[130] and it is not likely that it will ever appear.

Secondly, we have seen that incommensurability has an analogue in

130. 'The fact is that even totally different languages (like English and Hopi, or Chinese) are not untranslatable, and that there are many Hopis and Chinese who have learnt to master English very well', writes Popper ('Normal Science and its Dangers', *Criticism and the Growth of Knowledge*, op. cit., p. 56). He forgets that a proper translation has always had to do some violence either to English, or to the language translated. And who has ever denied that people can learn to move in mutually incommensurable frameworks? Exactly the same remark applies to Post's observation (p. 253 of his essay) that 'There are no barriers of communication between successive theories at least since the sixteenth century'. That possibility of communication does not entail comparability of meanings (commensurability) follows from this consideration (which I find in Körner's *Categorical Frameworks*, Oxford, 1971, p. 64): two persons, *A* and *B*, may speak two incommensurable languages *X* and *Y*. Yet *A*, interpreting every sentence of *B*'s as expressing a statement of *X* and *B* interpreting every sentence of *A* as expressing a statement of *Y*, may get on swimmingly with each other within a certain domain: 'Two propositions, *g* and *h*, may have a common informative content for *A* and *B* even if *g* is incompatible with the constitutive and individuating principles of *B*'s categorical framework and if *h* is incompatible with the constitutive and individuating principles of *A*'s categorical framework.' The reader is strongly advised

the field of perception and that it is part of the history of perception. This is the content of my *second thesis* on incommensurability: the development of perception and thought in the individual passes through stages which are mutually incommensurable.

My *third thesis* is that the views of scientists, and especially their views on basic matters, are often as different from each other as are the ideologies that underlie different cultures. Even worse: there exist scientific theories which are mutually incommensurable though they apparently deal 'with the same subject matter'. Not all competing theories have this property and those which have the property have it only as long as they are interpreted in a special way, for example, without reference to an 'independent observation language'. The illusion that we are dealing with the same subject matter arises in these cases as a result of an unconscious confusion of two different types of interpretation. Using an 'instrumentalistic' interpretation of the theories which sees in them no more than instruments for the classification of certain 'facts' one gets the impression that there is some common subject matter. Using a 'realistic' interpretation that tries to understand the theory in its own terms such a subject matter seems to disappear although there is the definite feeling (unconscious instrumentalism) that it must exist. Let us now examine how incommensurable theories might arise.

Scientific investigation, says Popper, *starts* with a problem and proceeds by *solving* it.

This characterization does not consider that problems may be wrongly formulated, that one may inquire about properties of things and processes which later views declare to be non-existent. Problems of this kind are not *solved*, they are *dissolved* and removed from the domain of legitimate inquiry. Examples are the problem of the absolute velocity of the earth, the problem of the trajectory of electrons in an interference pattern and the important problem of whether incubi are capable of producing offspring or whether they are forced to use the seeds of men for that purpose.[131]

---

to read Körner's book along with my own discussion of incommensurability (which is much more frustrating for logicians).

131. Cf. the *Malleus Malleficarum*, transl. Summers, London, 1928, Part II, Chapter IV, question 1. The theory goes back to St Thomas Aquinas.

The first problem was dissolved by the theory of relativity, which denies the existence of absolute velocities. The second problem was dissolved by the quantum theory, which denies the existence of trajectories in interference patterns. The third problem was dissolved, though less decisively, by modern (i.e. post-16th century) psychology and physiology, as well as by the mechanistic cosmology of Descartes.

Changes of ontology such as those just described are often accompanied by *conceptual changes*.

The discovery that certain entities do not exist may prompt the scientist to re-describe the events, processes, observations which were thought to be manifestations of them and which were therefore described in terms assuming their existence. (Or rather it may prompt him to introduce new *concepts*, since the older *words* will remain in use for a considerable time.) This applies especially to those 'discoveries' which suspend universal principles. The 'discovery' of an 'underlying substance' and of a 'spontaneous I' is of this kind, as we have seen.

An interesting development occurs when the faulty ontology is *comprehensive*, that is, when its elements are thought to be present in every process in a certain domain. In *this* case, *every* description inside the domain must be changed and must be replaced by a different statement (or by no statement at all). Classical physics is a case in point. It has developed a comprehensive terminology for describing some very fundamental properties of physical objects, such as shapes, masses, volumes, time intervals and so on. The conceptual system connected with this terminology assumes, at least in one of its numerous interpretations, that the properties *inhere* in objects and change only as the result of a direct physical interference. This is one of the 'universal principles' of classical physics. The theory of relativity implies, at least in the interpretation that was accepted by Einstein and Bohr, that inherent properties of the kind enumerated do not exist, that shapes, masses, time intervals are relations between physical objects and co-ordinate systems which may change, *without any physical interference*, when we replace one co-ordinate system by another. The theory of relativity also provides new principles for constituting mechanical facts. The new conceptual system that arises in this way does not just *deny* the existence of classical states of affairs, it does not even permit us to *formulate statements* expressing such states of affairs. It does not, and cannot, share a single statement with its predecessor – assuming all the time that we do not use the theories as classificatory schemes for the ordering of neutral facts. If we interpret both theories in a realistic manner, then the 'formal conditions for a suitable successor of a refuted theory', which were stated in Chapter 15 (it has to repeat the successful consequences of the older theory, deny its false consequences, and make additional predictions), cannot be satisfied and the positivistic scheme of progress with its 'Popperian spectacles', breaks down. Even Lakatos' liberalized version cannot survive this result; for it, too, assumes that content-classes of different theories can be compared, i.e. that a relation of inclusion, exclusion or overlap can be established between them. It is no use trying to connect classical statements with relativistic statements by an *empirical hypothesis*. A hypothesis of this kind would be as laughable as the statement 'whenever there is possession by a demon there is also discharge in the brain', which establishes a connection between terms of a possession theory of epilepsy and more recent 'scientific' terms. For we clearly do not want to perpetuate the older devilish terminology, and take it seriously, just in order to guarantee comparability of content-classes. But in the case of relativity vs. classical mechanics, a hypothesis of this kind *cannot even be formulated*. Using classical terms we assume a universal principle that is suspended by relativity which means it is suspended whenever we write down a sentence with the intention to express a relativistic state of affairs. Using classical terms and relativistic terms in the same statement we both use and suspend certain universal principles which is another way of saying that such statements do not exist: the case of relativity vs. classical mechanics is an example of two incommensurable frameworks. Other examples are the quantum theory vs. classical mechanics,[132] the

132. Bohr warns us (*Zs. Physik*, Vol. 13, 1922, p. 144) 'that the asymptotic connection' between the quantum theory and classical physics 'as it is assumed in the principle of correspondence . . . does not at all entail a gradual disappearance of the difference between the quantum theoretical treatment of radiation phenomena and the ideas of classical electrodynamics; all that is asserted is an asymptotic agreement of numerical statistical results.' In other words, the principle of correspondence asserts an agreement of *numbers*, not of *concepts*. According to Bohr, this agreement of numbers has even a certain disadvantage, for it '*obscures* the difference in principle between the laws which govern the actual mechanism of micro-processes and the continuous laws of the classical point of view' (p. 129; cf. also *Atomic Theory and the Description of*

impetus theory vs. Newton's mechanics,[133] materialism vs. mind-body dualism, and so on.

Now it is, of course, possible to interpret all these cases in a different way. Shapere, for example, has criticized my discussion of the impetus theory, saying that 'Newton himself is not perfectly clear as to whether inertial motion requires a cause'.[134] Moreover, he sees 'a large number of . . . resemblances and continuities' from Aristotle to Newton, where I see incommensurability.[135] The first objection can be removed quite easily (a) by pointing to the formulation of Newton's first law of motion – 'corpus omne persevarare *in statu quiescendi vel movendi* uniformiter in directum . . .' – which regards motion as a *state* rather than as a change;[136] (b) by showing that the notion of an impetus is defined in accordance with a law that is suspended by Newton and therefore ceases as a principle for constituting facts (this is done in some detail in my discussion of the case). Item (b) also takes care of the second objection: it is true that incommensurable frameworks and incommensurable concepts may exhibit many structural similarities – but this does not remove the fact that universal principles of the one framework are suspended by the other. It is *this* fact that establishes incommensurability despite all the similarities one might be able to discover.

Shapere (and others after him) have also tried to show that incommensurable theories are not just rare, but are a philosophical impossibility. I now turn to a discussion of these arguments.

Nature, Cambridge, 1932, pp. 85 and 87ff). Bohr has, therefore, repeatedly emphasized that 'the principle of correspondence must be regarded as a purely quantum theoretical law which cannot in any way diminish the contrast between the postulates [of the existence of stationary states and the transition postulate] and electro-magnetic theory' (ibid., p. 142, footnote). The difficulties arising from the neglect of this situation have been explained very clearly by the late N. R. Hanson in *Patterns of Discovery*, op. cit., Chapter 6; cf. also my comments in *Phil. Rev.*, Vol. 69, especially p. 251. They are not realized by flat-footed rationalists who infer continuity of concepts from the existence of approximations; cf. Popper's essay in *Criticism*, op. cit., p. 57.

133. Cf. my discussion of the impetus theory in *Minnesota Studies in the Philosophy of Science*, op. cit., Vol. 3.

134. 'Meaning and Scientific Change', in *Mind and Cosmos*, ed. Colodny, Pittsburgh, 1966, p. 78.

135. loc. cit.

136. Cf. A. Koyré, 'The Significance of the Newtonian Synthesis', in *Newtonian Studies*, London, 1965, pp. 9ff.

I have said that a scientific change may lead to a replacement of statements in a certain domain and that the replacement will be comprehensive when we are dealing with comprehensive ideologies. It will affect not only theories but also observational statements and (see Galileo above) natural interpretations. Now such an adaptation of observation to theory (and this is the gist of the *first objection*) often removes conflicting observation reports and saves a new cosmology in an *ad hoc* manner. Moreover, there arises the *suspicion* that observations which are interpreted in terms of a new theory can no longer be used to refute the theory. It is not difficult to reply to these points.

As regards the objection let me point out, in agreement with what has been said before (cf. Chapters 5 and 6), that an inconsistency between theory and observation may reveal a fault of the *observational terminology* (and even of our sensations), so that it is quite natural to change this terminology, adapt it to the new theory and see what happens. Such a change gives rise (and should give rise) to new auxiliary subjects (hydrodynamics, theory of solid objects, optics in the case of Galileo) which may more than compensate for the loss of empirical content. And as regards the suspicion,[137] we must remember that the predictions of a theory depend on its postulates (and associated grammatical rules) *and also* on initial conditions, while the meaning of the 'primitive' notions depends on the postulates (and associated grammatical rules) *only*. In those rare cases, however, where a theory entails assertions about possible initial conditions,[138] we can refute it with the help of *self-inconsistent observation reports* such as: 'object A does not move on a geodesic', which, if analysed in accordance with the Einstein–Infeld–Hoffmann account, reads 'singularity α which moves on a geodesic does not move on a geodesic'.

The *second objection* criticizes an interpretation of science that seems to be necessary for incommensurability to come about. I have already pointed out that the question 'are two particular comprehensive theories, such as classical mechanics and the special theory of relativity, incom-

137. The suspicion was voiced by Professor Hempel in a discussion at the Minnesota Center for the Philosophy of Science, cf. *Minnesota Studies*, Vol. IV, Minneapolis, 1970, pp. 236ff.

138. This seems to occur in some versions of the general theory of relativity, cf. Einstein–Infeld–Hoffmann, *Ann. Math.*, Vol. 39, 1938, p. 65, and Sen, op. cit., pp. 19ff.

mensurable?' is not a complete question. Theories can be interpreted in different ways. They will be commensurable in some interpretations, incommensurable in others. Instrumentalism, for example, makes commensurable all those theories which are related to the same observation language and are interpreted on its basis. A realist, on the other hand, wants to give a unified account, both of observable and of unobservable matters, and he will use the most abstract terms of whatever theory he is contemplating for that purpose.[139] He will use such terms in order either to *give* meaning to observation sentences or else to *replace* their customary interpretation. (For example, he will use the ideas of the special theory of relativity in order to replace the customary classical interpretation of everyday statements about shapes, temporal sequences, and so on.) Against this, it is pointed out by almost all empiricists that theoretical terms receive their interpretation from being connected with a pre-existing observation language, or with another theory that has already been connected with such a language. Thus Carnap asserts, in a passage I have already quoted,[140] that there is 'no independent interpretation for L_T [the language in terms of which a certain theory, or a certain world view, is formulated]. The system *T* [the axioms of the theory and the rules of derivation] is in itself an uninterpreted postulate system. [Its] terms . . . obtain only an indirect and incomplete interpretation by the fact that some of them are connected by the [correspondence] rules *C* with observation terms. . . .' Now, if theoretical terms have no 'independent interpretation', then they cannot be used for correcting the interpretation of observation statements which is the one and only source of their meaning. It follows that realism, as described here, is an impossible doctrine and that incommensurability cannot arise as long as we keep within the confines of 'sound' (i.e. empiristic) scientific method.

139. This consideration has been raised into a principle by Bohr and Rosenfeld, *Kgl. Danske Videnskab, Selskab Mat.-Phs. Medd.*, Vol. 12, No. 8, 1933, and, more recently, by Marzke and Wheeler in 'Gravitation and Geometry I', op. cit., p. 48: 'Every proper theory should provide, in and by itself, its own means for defining the quantities with which it deals. According to this principle, classical general relativity should admit to calibrations of space and time that are altogether free of any reference to [objects which are external] to it such as rigid rods, inertial clocks, or atomic clocks [which involve] the quantum of action.' Its terms should also be free of contamination by observational concepts that belong to an earlier, and more primitive, stage of knowledge.

140. See footnote 13 to Chapter 12.

The guiding idea behind this very popular objection is that new and abstract languages cannot be introduced in a direct way, but must first be connected with an already existing, and presumably stable, observation idiom.[141]

This guiding idea is refuted at once by noting the way in which children learn to speak – they certainly do not start from an innate observation language – and the way in which anthropologists and linguists learn the unknown language of a newly-discovered tribe.

The first case has already been briefly described. In the second case we see that what is anathema in anthropology, and with very good reasons, is still a fundamental principle for the contemporary representatives of the philosophy of the Vienna circle. According to Carnap, Feigl, Hempel, Nagel and others the terms of a theory receive their interpretation in an indirect fashion, by being related to a different conceptual system which is either an older theory or an observation language.[142]

141. An even more conservative principle is sometimes used when discussing the possibility of languages with a logic different from our own: 'Any allegedly new possibility must be capable of being fitted into, or understood in terms of, our present conceptual or linguistic apparatus.' B. Stroud, 'Conventionalism and the Indeterminacy of Translations', *Synthèse*, 1968, p. 173. The idea that a new language must be taught with the help of a stable observation language derives from the empirical tradition, and especially from the views on *logical reconstruction* which arose in the Vienna Circle. According to these views, the empirical content of a theory (or of some common idiom) is found by inquiring how much of the theory (or of the idiom) can be translated into an *ideal language* whose empirical characteristics can be easily ascertained. Theories are regarded as meaningful only to the extent to which such a translation can be carried out. On the basis of these views, it was natural to try tying the acquisition of a new language to the chosen ideal language. However, it soon turned out that ideal languages are not easy to come by and that even the most elementary first steps are beset by problems entirely absent from physics. In addition, the concepts of the chosen 'bases' of reconstruction had to be gradually enriched to cope with the intersubjectivity of scientific terms. A series of developments of this kind, few of them clearly understood *or even recognized*, then led to the gradual replacement of the idea of reconstruction by the idea of interpretation and from there to the idea of teaching (cf. Hempel's perceptive criticism of the present situation in *Minnesota Studies*, Vol. 4, Minneapolis, 1970, pp. 162ff). The development, with its mistakes and oversights, with its gradual sliding from one position into another, is unknown to J. Giedymin (*BJPS*, 22, 1971, pp. 40ff) who criticizes me for taking it into account. Giedymin's ignorance is not surprising, for he explicitly refuses to take history seriously (*BJPS*, 21, 1970, p. 257).

142. For what follows, cf. also my review of Nagel's *Structure of Science*, in *British Journal for the Philosophy of Science*, Vol. 6, 1966, pp. 237–49.

Older theories or observation languages are adopted not because of their theoretical excellence (they cannot possibly be – the older theories are usually refuted). They are adopted because they are 'used by a certain language community as a means of communication'.[143] According to this method the phrase 'having much larger relativistic mass than . . .' is partially interpreted by first connecting it with some *pre-relativistic terms* (classical terms, common-sense terms) which are 'commonly understood' (presumably, as the result of previous teaching in connection with crude weighing methods), and it is used only after such connection has given it a more or less definite content.

This procedure, whose application may involve a formidable logical apparatus and which is therefore often regarded as the *dernier cri* of a truly scientific philosophy, is even worse than the once quite popular demand to clarify doubtful points by translating them into Latin. For while Latin was chosen because of its precision and clarity, and also because it was conceptually richer than the slowly evolving vulgar idioms,[144] while it was chosen for a theoretical reason, the choice of an observation language or of an older theory, is due to the fact that they are 'antecedently understood': it is due to their *popularity*. Besides, if pre-relativistic terms, which are pretty far removed from reality (especially in view of the fact that they come from an incorrect theory based on non-existent ontology), can be taught ostensively, for example, with the help of crude weighing methods (and we must assume that they can be so taught or the whole scheme collapses at once), then why should we not introduce the relativistic terms *directly* and *without* assistance from the terms of some other idiom? Finally, it is plain common sense that the teaching, or the learning, or the constructing, of new and unknown languages must not be contaminated by extraneous material. Linguists remind us that a perfect translation is never possible, even if we use

143. Carnap, op. cit., p. 40; cf. also Hempel, *Philosophy of Natural Science*, New York, 1966, pp. 74ff.

144. It was for this reason that Leibniz regarded the German of his time, and especially the German of the artisans, as a perfect observation language, while Latin for him was already over-contaminated by theoretical notions. See his 'Unvergreifliche Gedancken, betreffend die Ausuebung und Verbesserung der Teutschen Sprache', *Wissenschaftliche Beihefte zur Zeitschrift des allgemeinen deutschen Sprachvereins*, IV. Reihe, Heft 29, Berlin, 1907, pp. 292ff.

complex contextual definitions. This is one of the reasons for the importance of *field work*, where new languages are learned *from scratch*, and for the rejection, as inadequate, of any account that relies on a complete or a partial translation. *Yet just what is anathema in linguistics is now taken for granted by logical empiricists*, a mythical 'observation language' replacing the English of the translators. Let us commence field work in this domain also, and let us study the language of new theories not in the definition-factories of the double language model, but in the company of those metaphysicians, physicists, playwrights, courtesans, who have constructed new world views! This finishes my discussion of the guiding principle behind the second objection against realism and the possibility of *incommensurable* theories.

A *third objection* is that there exist *crucial experiments* which refute one of two allegedly incommensurable theories and confirm the other. For example, the Michelson–Morley experiment, the variation of the mass of elementary particles, the transverse Doppler effect, are said to refute classical mechanics and to confirm relativity. The answer to this problem is not difficult either. Adopting the point of view of relativity we find that the experiments, *which of course will now be described in relativistic terms*, using the relativistic notions of length, duration, mass, speed, and so on,[145] are *relevant* to the theory, and we also find that they *support* the theory. Adopting classical mechanics (with, or without, an ether) we again find that the experiments *which are now described in the very different terms of classical physics* (i.e. roughly in the manner in which Lorentz described them) are relevant, but we also find that they *undermine* (the conjunction of electrodynamics and) classical mechanics. Why should it be necessary to possess terminology that allows us to say that it is the same experiment which confirms one theory and refutes the other? But did we not ourselves use such terminology? Well, for one thing it should be easy, though somewhat laborious to express what was just said *without* assuming identity. Secondly, the identification is, of course, not contrary to my thesis, for we are now not *using* the terms of

145. For examples of such descriptions cf. Synge, 'Introduction to General Relativity', Section II, in *Relativity, Groups, and Topology*, ed. de Witt and de Witt, 1964. For an even more elegant way of introducing relativity cf. Bondi, *Assumption and Myth in Physical Theory*, Cambridge, 1967, pp. 29ff (the *k*-calculus).

either relativity or of classical physics, as is done in a test, but are *referring to* them and their relation to the physical world. The language in which *this* discourse is carried out can be classical, or relativistic, or Voodoo. It is no good insisting that scientists act as if the situation were much less complicated.[146] If they act that way, then they are either instrumentalists (see above) or mistaken: many scientists are nowadays interested in *formulae*, while I am discussing *interpretations*. It is also possible that being well acquainted with both theories they change back and forth between them with such speed that they seem to remain within a single domain of discourse.

(This last remark, incidentally, takes care also of the objection that 'the transition from Newton's theory of gravity to Einstein's cannot be an irrational leap' because Newton's theory 'follows from Einstein's theory' as an excellent approximation.[147] Good thinkers can leap rather quickly and continuity of formal relations does not entail continuity of interpretations, as anyone familiar with the notorious 'derivation' of the law of gravitation from Kepler's laws is by now bound to know.)

It is also said that by admitting incommensurability into science we can no longer decide whether a new view *explains* what it is supposed to explain, or whether it does not wander off into different fields.[148] For example, we would not know whether a newly invented physical theory is still dealing with problems of space and time, or whether its author has not by mistake made a biological assertion. But there is no need to possess such knowledge. For once the fact of incommensurability has been admitted, the question which underlies the objection does not arise (conceptual progress often makes it impossible to ask certain questions and to explain certain things; thus, we can no longer ask for the absolute velocity of an object, at least as long as we take relativity seriously). Is this a serious loss for science? Not at all! Progress was made by the very same 'wandering off into different fields' whose undecidability now so greatly exercises the critic: Aristotle saw the world as a super-*organism*, as a *biological* entity, while one essential element of the new science of

146. As does Popper, op. cit., p. 57.

147. Popper, op. cit.

148. This difficulty was raised by Roger Buck in a discussion at the Minnesota Center, cf. *Minnesota Studies*, Vol. 4, p. 232.

Descartes, Galileo and their followers in medicine and biology is its exclusively mechanistic outlook. Are such developments to be forbidden? And if they are not, what, then, is left of the complaint?

A closely connected objection starts from the notion of *explanation* or *reduction* and emphasizes that this notion assumes continuity of concepts (other notions could be used for starting exactly the same kind of argument). To take my example, relativity is supposed to explain the valid parts of classical physics, hence it cannot be incommensurable with it! The reply is again obvious! Why should the relativist be concerned with the fate of classical mechanics except as part of a historical exercise? There is only *one* task we can legitimately demand of a theory, and it is that it should give us a correct account of the world, i.e. of the totality of facts *as constituted by its own basic concepts*. What have the principles of explanation got to do with this demand? Is it not reasonable to assume that a point of view, such as the point of view of classical mechanics, that has been found wanting in various respects and that gets into difficulty *with its own facts* (see above, on crucial experiments), cannot have entirely adequate concepts? Is it not equally reasonable to try to replace its concepts by those of a more successful cosmology? Besides, why should the notion of explanation be burdened by the demand for conceptual continuity? This notion has been found to be too narrow before (demand for derivability) and it had to be widened to include partial and statistical connections. Nothing prevents us from widening it still further to admit, say, 'explanations by equivocation'.

Incommensurable theories, then, can be *refuted* by reference to their own respective kinds of experience; i.e. by discovering the *internal contradictions* from which they are suffering. (In the absence of commensurable alternatives these refutations are quite weak, however, as can be seen from the arguments for proliferation in Chapters 2 and 3.) Their *contents* cannot be compared. Nor is it possible to make a judgement of *verisimilitude* except within the confines of a particular theory (remember that the problem of incommensurability arises only when we analyse the change of *comprehensive cosmological points of view* – restricted theories rarely lead to the needed conceptual revisions). None of the methods which Carnap, Hempel, Nagel, Popper or even Lakatos want to use for rationalizing scientific changes can be applied, and the one that *can* be applied, refutation, is greatly reduced in strength. What remains are aesthetic judgements, judgements of taste, metaphysical prejudices, religious desires, in short, *what remains are our subjective wishes*: science at its most advanced and general returns to the individual a freedom he seems to lose when entering its more pedestrian parts, and even its 'third world' image, the development of its concepts, ceases to be 'rational'. This is the last argument needed for retaining the conclusion of Chapter 15 (and of the entire book) despite the attacks of our most modern and most sophisticated rationalists.

Appendix 5

Whorff speaks of 'Ideas', not of 'events' or of 'facts', and it is not always clear whether he would approve of my extension of his views. On the one hand he says that 'time, velocity, and matter are not essential to the construction of a consistent picture of the universe' (p. 216), and he asserts that 'we cut up nature, organise it into concepts, and ascribe significances as we do, largely because we are partial to an agreement to organize it in this way' (p. 213), which would seem to imply that widely different languages posit not just different ideas for the ordering of the same facts, but that they posit also different facts. The 'linguistic relativity principle' seems to point in the same direction. It says, 'in informal terms, that users of markedly different grammars are pointed by their grammars towards different types of observations and different evaluations of externally similar acts of observation, and hence are not equivalent observers, but must arrive at somewhat different views of the world' (p. 221). But the 'more formal statements' (p. 221) of the principle already contains a different element, for here we are told that 'all observers are not led by *the same physical evidence* to the same picture of the universe, unless their linguistic backgrounds are similar, or can in some way be calibrated' (p. 214, my italics), which can either mean that observers using widely different languages will *posit different facts* under the same physical circumstances in the same physical world, or it can mean that they will *arrange similar facts in different ways*. The second interpretation finds some support in the examples given, where different isolates of meaning in English and Shawnee are said to be 'used in reporting *the same experience*' (p. 208) and where we read that 'languages classify items of experience differently' (p. 209); experience is regarded as a uniform reservoir of facts which are *classified* differently by different languages. It finds further support in Whorff's description of the transition from the *horror-vacui* account of barometric phenomena to the modern theory: 'If once these sentences [Why does water rise in a pump? Because Nature abhors a vacuum.] seemed satisfying to logic, but today seem idiosyncrasies of a particular jargon, the change did not come about because science has discovered new facts. Science has adopted new linguistic formulations of the old facts, and now that we have become at home in the new dialect, certain traits of the old one are no longer binding on us' (p. 222). However, I regard these more conservative statements as secondary when compared with the great influence ascribed to grammatical categories and especially to the more hidden 'rapport systems' of a language (pp. 68ff).

Some philosophers might want to connect incommensurability with the issues raised by what has been called 'radical translation'. As far as I can see this is not going to advance matters. Radical translation is a triviality blown up into a major philosophical discovery: neither behaviour nor observational data of a more subjective kind can ever determine interpretations (for this point cf. my paper 'Towards a Realistic Interpretation of Experience', *Proceedings of the Aristotelian Society*, 1958). And the further ramification of this triviality (such as Davidson's refrigerated hippopotamus) could occur only because linguistic philosophers do not seem to be familiar with the problems, the techniques, and the objections to conventionalism. Besides, *our* problem is one of historical fact and *not* of logical possibility.

Illustration A. Ajax and Achilles, playing at dice. Vatican Museum.
(Courtesy the Vatican Museum)

Illustration B. Vase from the Dipylon Cemetery, Athens. Mid 8th Century.
(Courtesy the Mansell Collection)

Illustration C. Attic Funerary Crater, around 750 B.C.
National Museum, Athens. (Courtesy the Mansell Collection)

Illustration D. Warrior Vase, Mycenae acropolis, around 1200 B.C. Athens, National Museum. (Courtesy the Mansell Collection)

Illustration E. Detail from Illustration C: Charioteers with chariots.

18

Thus science is much closer to myth than a scientific philosophy is prepared to admit. It is one of the many forms of thought that have been developed by man, and not necessarily the best. It is conspicuous, noisy, and impudent, but it is inherently superior only for those who have already decided in favour of a certain ideology, or who have accepted it without ever having examined its advantages and its limits. And as the accepting and rejecting of ideologies should be left to the individual it follows that the separation of state and church *must be complemented by the separation of state and* science, *that most recent, most aggressive, and most dogmatic religious institution. Such a separation may be our only chance to achieve a humanity we are capable of, but have never fully realized.*

The idea that science can, and should, be run according to fixed and universal rules, is both unrealistic and pernicious. It is *unrealistic*, for it takes too simple a view of the talents of man and of the circumstances which encourage, or cause, their development. And it is *pernicious*, for the attempt to enforce the rules is bound to increase our professional qualifications at the expense of our humanity. In addition, the idea is *detrimental to science*, for it neglects the complex physical and historical conditions which influence scientific change. It makes our science less adaptable and more dogmatic: every methodological rule is associated with cosmological assumptions, so that using the rule we take it for granted that the assumptions are correct. Naive falsificationism takes it for granted that the laws of nature are manifest and not hidden beneath disturbances of considerable magnitude. Empiricism takes it for granted that sense experience is a better mirror of the world than pure thought. Praise of argument takes it for granted that the artifices of Reason give better results than the unchecked play of our emotions. Such assumptions may be perfectly plausible *and even true*. Still, one should occasionally put them to a test. Putting them to a test means that we stop using the methodology associated with them, start doing science in a different way and see what happens. Case studies such as those reported in the preceding chapters show that such tests occur all the time, and that they speak *against* the universal validity of any rule. All methodologies have their limitations and the only 'rule' that survives is 'anything goes'.

The change of perspective brought about by these discoveries leads once more to the long-forgotten problem of the excellence of science. It leads to it for the first time in *modern* history, for modern science *overpowered* its opponents, it did not *convince* them. Science took over by *force*, not by argument (this is especially true of the former colonies where science and the religion of brotherly love were introduced as a matter of course, and without consulting, or arguing with, the inhabitants). Today we realize that rationalism, being bound to science, cannot give us any assistance in the issue between science and myth and we also know, from inquiries of an entirely different kind, that myths are vastly better than rationalists have dared to admit.[1] Thus we are now *forced* to raise the question of the excellence of science. An examination then reveals that science and myth overlap in many ways, that the differences we think we perceive are often *local* phenomena which may turn into similarities elsewhere and that fundamental discrepancies are results of different *aims* rather than of different methods trying to reach one and the same 'rational' end (such as, for example, 'progress', or increase of content, or 'growth').

To show the surprising similarities of myth and science, I shall briefly discuss an interesting paper by Robin Horton, entitled 'African Traditional Thought and Western Science'.[2] Horton examines African

1. Cf. the marvellous case studies by Evans-Pritchard, Griaule, Edith Hamilton, Jeremias, Frankfort, Thorkild Jacobsen and others. For a survey cf. de Santillana-von Dechend, *Hamlet's Mill*, Boston, 1969, as well as my *Einführung in die Naturphilosophie*, Braunschweig, 1976. These are case studies in the sense of Lakatos and they satisfy his most stringent criteria. Why, then, are he and his fellow-rationalists so reluctant to accept their results?

2. Originally published in *Africa*, Vol. 37, 1967, pp. 87–155. I am quoting from the

mythology and discovers the following features: the quest for theory is a quest for unity underlying apparent complexity. The theory places things in a causal context that is wider than the causal context provided by common sense: both science and myth cap common sense with a theoretical superstructure. There are theories of different degrees of abstraction and they are used in accordance with the different requirements of explanation that arise. Theory construction consists in breaking up objects of common sense and in reuniting the elements in a different way. Theoretical models start from analogy but they gradually move away from the pattern on which the analogy was based. And so on.

These features, which emerge from case studies no less careful and detailed than those of Lakatos, refute the assumption that science and myth obey different principles of formation (Cassirer), that myth proceeds without reflection (Dardel), or speculation (Frankfort, occasionally). Nor can we accept the idea, found in Malinowski but also in classical scholars such as Harrison and Cornford, that myth has an essentially pragmatic function or is based on ritual. Myth is much closer to science than one would expect from a philosophical discussion. It is closer to science than even Horton himself is prepared to admit.

To see this, consider some of the *differences* Horton emphasizes. According to Horton, the central ideas of a myth are regarded as sacred. There is anxiety about threats to them. One 'almost never finds a confession of ignorance'[3] and events 'which seriously defy the established lines of classification in the culture where they occur' evoke a 'taboo reaction'.[4] Basic beliefs are protected by this reaction as well as by the device of 'secondary elaborations'[5] which, in our terms, are series of *ad hoc* hypotheses. Science, on the other hand, is characterized by an 'essential scepticism'[6]; 'when failures start to come thick and fast, defence of the theory switches inexorably to attack on it'.[7] This is possible because of the 'openness' of the scientific enterprise, because of the pluralism of ideas it contains and also because 'whatever defies or fails to fit into the established category system is not something horrifying, to

---

abbreviated reprint in Max Marwick (ed.), *Witchcraft and Sorcery*, Penguin Books, 1970, pp. 342ff.

3. ibid., p. 362.
4. ibid., p. 364.
5. ibid., p. 365.
6. ibid., p. 358.
7. loc. cit.

be isolated or expelled. On the contrary, it is an intriguing 'phenomenon' – a starting-point and a challenge for the invention of new classifications and new theories'.[8] We can see that Horton has read his Popper well.[9] A field study of science itself shows a very different picture.

Such a study reveals that, while some scientists may proceed as described, the great majority follow a different path. Scepticism is at a minimum; it is directed against the view of the opposition and against minor ramifications of one's own basic ideas, never against the basic ideas themselves.[10] Attacking the basic ideas evokes taboo reactions which are no weaker than are the taboo reactions in so-called primitive societies.[11] Basic beliefs are protected by this reaction as well as by secondary elaborations, as we have seen, and whatever fails to fit into the established category system or is said to be incompatible with this system is either viewed as something quite horrifying or, more frequently, *it is simply declared to be non-existent*. Nor is science prepared to make a theoretical pluralism the foundation of research. Newton reigned for more than 150 years, Einstein briefly introduced a more liberal point of view only to be succeeded by the Copenhagen Interpretation. The similarities between science and myth are indeed astonishing.

But the fields are even more closely related. The massive dogmatism I have described is not just a *fact*, it has also a most important *function*. *Science would be impossible without it*.[12] 'Primitive' thinkers showed greater insight into the nature of knowledge than their 'enlightened' philosophical rivals. It is, therefore, necessary to re-examine our attitude towards myth, religion, magic, witchcraft and towards all those ideas which rationalists would like to see forever removed from the surface of

8. ibid., p. 365.

9. See his discussion of what he calls the 'Closed and Open Predicament' in Part 2 of his essay.

10. This is a very familiar procedure in African witchcraft. Cf. Evans-Pritchard, *Witchcraft, Oracles and Magic Among the Azande*, Oxford, 1937, pp. 230, 338; also *Social Anthropology*, op. cit., p. 99.

11. Cf. the early reactions against hidden variables in the quantum theory, the attitude towards astrology, telekinesis, telepathy, Voodoo, Ehrenhaft, Velikovsky, and so on. Cf. also Köstler's amusing story *The Midwife Toad*, New York, 1973.

12. This has been emphasized by Kuhn; see 'The Function of Dogma in Scientific Research' in A. C. Crombie (ed.), *Scientific Change*, London, 1963, pp. 69–347, and *The Structure of Scientific Revolutions*, Chicago, 1962.

the earth (without having so much as looked at them – a typical taboo reaction).

There is another reason why such a re-examination is urgently required. The rise of modern science coincides with the suppression of non-Western tribes by Western invaders. The tribes are not only physically suppressed, they also lose their intellectual independence and are forced to adopt the bloodthirsty religion of brotherly love – Christianity. The most intelligent members get an extra bonus: they are introduced into the mysteries of Western Rationalism and its peak – Western Science. Occasionally this leads to an almost unbearable tension with tradition (Haiti). In most cases the tradition disappears without the trace of an argument, one simply becomes a slave both in body and in mind. Today this development is gradually reversed – with great reluctance, to be sure, but it is reversed. Freedom is regained, old traditions are rediscovered, both among the minorities in Western countries and among large populations in non-Western continents. *But science still reigns supreme*. It reigns supreme because its practitioners are *unable to understand*, and *unwilling to condone*, different ideologies, because they have the *power* to enforce their wishes, and because they *use* this power just as their ancestors used *their* power to force Christianity on the peoples they encountered during their conquests. Thus, while an American can now choose the religion he likes, he is still not permitted to demand that his children learn magic rather than science at school. There is a separation between state and church, there is no separation between state and science.

And yet science has no greater authority than any other form of life. Its aims are certainly not more important than are the aims that guide the lives in a religious community or in a tribe that is united by a myth. At any rate, they have no business restricting the lives, the thoughts, the education of the members of a free society where everyone should have a chance to make up his own mind and to live in accordance with the social beliefs he finds most acceptable. The separation between state and church must therefore be complemented by the separation between state and science.

We need not fear that such a separation will lead to a breakdown of technology. There will always be people who prefer being scientists to being the masters of their fate and who gladly submit to the meanest kind of (intellectual and institutional) slavery provided they are paid well and provided also there are some people around who examine their work and sing their praise. Greece developed and progressed because it could rely on the services of unwilling slaves. We shall develop and progress with the help of the numerous *willing* slaves in universities and laboratories who provide us with pills, gas, electricity, atom bombs, frozen dinners and, occasionally, with a few interesting fairy-tales. We shall treat these slaves well, we shall even listen to them, for they have occasionally some interesting stories to tell, but we shall *not* permit them to impose their ideology on our children in the guise of 'progressive' theories of education.[13] We shall not permit them to teach the fancies of science as if they were the only factual statements in existence. This separation of science and state may be our only chance to overcome the hectic barbarism of our scientific-technical age and to achieve a humanity we are capable of, but have never fully realized.[14] Let us, therefore, in conclusion review the arguments that can be adduced for such a procedure.

The image of 20th-century science in the minds of scientists and laymen is determined by technological miracles such as colour television, the moon shots, the infra-red oven, as well as by a somewhat vague but still quite influential rumour, or fairy-tale, concerning the manner in which these miracles are produced.

According to the fairy-tale the success of science is the result of a subtle, but carefully balanced combination of inventiveness and control. Scientists have *ideas*. And they have special *methods* for improving ideas. The theories of science have passed the test of method. They give a better account of the world than ideas which have not passed the test.

The fairy-tale explains why modern society treats science in a special way and why it grants it privileges not enjoyed by other institutions.

Ideally, the modern state is ideologically neutral. Religion, myth, prejudices *do* have an influence, but only in a roundabout way, through the medium of politically influential *parties*. Ideological principles *may*

13. Cf. Appendix 3, p. 215.

14. For the humanitarian deficiencies of science cf. 'Experts in a Free Society', *The Critic*, November/December 1971, or the improved German version of this essay and of 'Towards a Humanitarian Science' in Part II of Vol. I of my *Ausgewählte Aufsätze*. Vieweg, 1974.

enter the governmental structure, but only via a majority vote, and after a lengthy discussion of possible consequences. In our schools the main religions are taught as *historical phenomena*. They are taught as parts of the truth only if the parents insist on a more direct mode of instruction. It is up to them to decide about the religious education of their children. The financial support of ideologies does not exceed the financial support granted to parties and to private groups. State and ideology, state and church, state and myth, are carefully separated.

State and science, however, work closely together. Immense sums are spent on the improvement of scientific ideas. Bastard subjects such as the philosophy of science which have not a single discovery to their credit profit from the boom of the sciences. Even human relations are dealt with in a scientific manner, as is shown by education programmes, proposals for prison reform, army training, and so on. Almost all scientific subjects are compulsory subjects in our schools. While the parents of a six-year-old child can decide to have him instructed in the rudiments of Protestantism, or in the rudiments of the Jewish faith, or to omit religious instruction altogether, they do not have a similar freedom in the case of the sciences. Physics, astronomy, history *must* be learned. They cannot be replaced by magic, astrology, or by a study of legends.

Nor is one content with a merely *historical* presentation of physical (astronomical, historical, etc.) facts and principles. One does not say: *some people believe* that the earth moves round the sun while others regard the earth as a hollow sphere that contains the sun, the planets, the fixed stars. One says: the earth *moves* round the sun – everything else is sheer idiocy.

Finally, the manner in which we accept or reject scientific ideas is radically different from democratic decision procedures. We accept scientific laws and scientific facts, we teach them in our schools, we make them the basis of important political decisions, but without ever having subjected them to a vote. *Scientists* do not subject them to a vote – or at least this is what they say – and *laymen* certainly do not subject them to a vote. Concrete proposals are occasionally discussed, and a vote is suggested. But the procedure is not extended to general theories and scientific facts. Modern society is 'Copernican' not because Copernicanism has been put on a ballot, subjected to a democratic debate and then voted in with a simple majority; it is 'Copernican' because the *scientists* are Copernicans and because one accepts their cosmology as uncritically as one once accepted the cosmology of bishops and cardinals.

Even bold and revolutionary thinkers bow to the judgement of science. Kropotkin wants to break up all existing institutions – but he does not touch science. Ibsen goes very far in unmasking the conditions of contemporary humanity – but he still retains science as a measure of the truth. Evans-Pritchard, Lévi-Strauss and others have recognized that 'Western Thought', far from being a lonely peak of human development, is troubled by problems not found in other ideologies – but they exclude science from their relativization of all forms of thought. Even for them science is a *neutral structure* containing *positive knowledge* that is independent of culture, ideology, prejudice.

The reason for this special treatment of science is, of course, our little fairy-tale: if science has found a method that turns ideologically contaminated ideas into true and useful theories, then it is indeed not mere ideology, but an objective measure of all ideologies. It is then not subjected to the demand for a separation between state and ideology.

But the fairy-tale is false, as we have seen. There is no special method that guarantees success or makes it probable. Scientists do not solve problems because they possess a magic wand – methodology, or a theory of rationality – but because they have studied a problem for a long time, because they know the situation fairly well, because they are not too dumb (though that is rather doubtful nowadays when almost anyone can become a scientist), and because the excesses of one scientific school are almost always balanced by the excesses of some other school. (Besides, scientists only rarely solve their problems, they make lots of mistakes, and many of their solutions are quite useless.) Basically there is hardly any difference between the process that leads to the announcement of a new scientific law and the process preceding passage of a new law in society: one informs either all citizens or those immediately concerned, one collects 'facts' and prejudices, one discusses the matter, and one finally votes. But while a democracy makes some effort to *explain* the process so that everyone can understand it, scientists either *conceal* it, or *bend* it, to make it fit their sectarian interests.

No scientist will admit that voting plays a role in his subject. Facts, logic, and methodology alone decide – this is what the fairy-tale tells us. But how do facts decide? What is their function in the advancement of knowledge? We cannot *derive* our theories from them. We cannot give a *negative* criterion by saying, for example, that good theories are theories which can be refuted, but which are not yet contradicted by any fact. A principle of falsification that removes theories because they do not fit the facts would have to remove the whole of science (or it would have to admit that large parts of science are irrefutable). The hint that a good theory *explains more* than its rivals is not very realistic either. True: new theories often predict new things – but almost always at the expense of things already known. Turning to logic we realize that even the simplest demands *are not* satisfied in scientific practice, and *could not be* satisfied, because of the complexity of the material. The ideas which scientists use to present the known and to advance into the unknown are only rarely in agreement with the strict injunctions of logic or pure mathematics and the attempt to make them conform would rob science of the elasticity without which progress cannot be achieved. We see: facts alone are not strong enough for making us accept, or reject, scientific theories, the range they leave to thought is *too wide*; logic and methodology eliminate too much, they are *too narrow*. In between these two extremes lies the ever-changing domain of human ideas and wishes. And a more detailed analysis of successful moves in the game of science ('successful' from the point of view of the scientists themselves) shows indeed that there is a wide range of freedom that *demands* a multiplicity of ideas and *permits* the application of democratic procedures (ballot-discussion-vote) but that is actually closed by power politics and propaganda. *This is where the fairy-tale of a special method assumes its decisive function.* It conceals the freedom of decision which creative scientists and the general public have even inside the most rigid and the most advanced parts of science by a recitation of 'objective' criteria and it thus protects the big-shots (Nobel Prize winners; heads of laboratories, of organizations such as the AMA, of special schools; 'educators'; etc.) from the masses (laymen; experts in non-scientific fields; experts in other fields of science): only those citizens count who were subjected to the pressures of scientific institutions (they have undergone a long process of education), who succumbed to these pressures (they have passed their examinations), and who are now firmly convinced of the truth of the fairy-tale. This is how scientists have deceived themselves and everyone else about their business, but without any real disadvantage: they have more money, more authority, more sex appeal than they deserve, and the most stupid procedures and the most laughable results in their domain are surrounded with an aura of excellence. It is time to cut them down in size, and to give them a more modest position in society.

This advice, which only few of our well-conditioned contemporaries are prepared to accept, seems to clash with certain simple and widely-known facts.

Is it not a fact that a learned physician is better equipped to diagnose and to cure an illness than a layman or the medicine-man of a primitive society? Is it not a fact that epidemics and dangerous individual diseases have disappeared only with the beginning of modern medicine? Must we not admit that technology has made tremendous advances since the rise of modern science? And are not the moon-shots a most impressive and undeniable proof of its excellence? These are some of the questions which are thrown at the impudent wretch who dares to criticize the special position of the sciences.

The questions reach their polemical aim only if one assumes that the results of science *which no one will deny* have arisen without any help from non-scientific elements, and that they cannot be improved by an admixture of such elements either. 'Unscientific' procedures such as the herbal lore of witches and cunning men, the astronomy of mystics, the treatment of the ill in primitive societies are totally without merit. *Science alone* gives us a useful astronomy, an effective medicine, a trustworthy technology. One must also assume that science owes its success to the correct method and not merely to a lucky accident. It was not a fortunate cosmological guess that led to progress, but the correct *and cosmologically neutral* handling of data. These are the assumptions we must make to give the questions the polemical force they are supposed to have. Not a single one of them stands up to closer examination.

Modern astronomy started with the attempt of Copernicus to adapt the old ideas of Philolaos to the needs of astronomical predictions. Philolaos was not a precise scientist, he was a muddleheaded Pythagorean, as we have seen (Chapter 5, footnote 25), and the consequences of his doctrine were called 'incredibly ridiculous' by a professional astronomer such as Ptolemy (Chapter 4, footnote 4). Even Galileo, who had the much improved Copernican version of Philolaos before him, says: 'There is no limit to my astonishment when I reflect that Aristarchus and Copernicus were able to make reason to conquer sense that, in defiance of the latter, the former became mistress of their belief' (*Dialogue*, 328). 'Sense' here refers to the experiences which Aristotle and others had used to show that the earth must be at rest. The 'reason' which Copernicus opposes to their arguments is the very mystical reason of Philolaos combined with an equally mystical faith ('mystical' from the point of view of today's rationalists) in the fundamental character of circular motion. I have shown that modern astronomy and modern dynamics could not have advanced without this unscientific use of antediluvian ideas.

While astronomy profited from Pythagoreanism and from the Platonic love for circles, medicine profited from herbalism, from the psychology, the metaphysics, the physiology of witches, midwives, cunning men, wandering druggists. It is well known that 16th- and 17th-century medicine while theoretically hypertrophic was quite helpless in the face of disease (and stayed that way for a long time after the 'scientific revolution'). Innovators such as Paracelsus fell back on the earlier ideas and improved medicine. Everywhere science is enriched by unscientific methods and unscientific results, while procedures which have often been regarded as essential parts of science are quietly suspended or circumvented.

The process is not restricted to the early history of modern science. It is not merely a consequence of the primitive state of the sciences of the 16th and 17th centuries. Even today science can and does profit from an admixture of unscientific ingredients. An example which was discussed above, in Chapter 4, is the revival of traditional medicine in Communist China. When the Communists in the fifties forced hospitals and medical schools to teach the ideas and the methods contained in the *Yellow Emperor's Textbook of Internal Medicine* and to use them in the treatment of patients, many Western experts (among them Eccles, one of the 'Popperian Knights') were aghast and predicted the downfall of Chinese medicine. What happened was the exact opposite. Acupuncture, moxibustion, pulse diagnosis have led to new insights, new methods of treatment, new problems both for the Western and for the Chinese physician.

And those who do not like to see the state meddling in scientific matters should remember the sizeable chauvinism of science: for most scientists the slogan 'freedom for science' means the freedom to indoctrinate not only those who have joined them, but the rest of society as well. Of course – not every mixture of scientific and non-scientific elements is successful (example: Lysenko). But science is not always successful either. If mixtures are to be avoided because they occasionally misfire, then pure science (if there is such a thing) must be avoided as well. (It is not the *interference* of the state that is objectionable in the Lysenko case, but the *totalitarian* interference that kills the opponent instead of letting him go his own way.)

Combining this observation with the insight that science has no special method, we arrive at the result that the separation of science and non-science is not only artificial but also detrimental to the advancement of knowledge. If we want to understand nature, if we want to master our physical surroundings, then we must use *all* ideas, *all* methods, and not just a small selection of them. The assertion, however, that there is no knowledge outside science – *extra scientiam nulla salus* – is nothing but another and most convenient fairy-tale. Primitive tribes have more detailed classifications of animals and plants than contemporary scientific zoology and botany, they know remedies whose effectiveness astounds physicians (while the pharmaceutical industry already smells here a new source of income), they have means of influencing their fellow men which science for a long time regarded as non-existent (Voodoo), they solve difficult problems in ways which are still not quite understood (building of the pyramids; Polynesian travels), there existed a highly developed and internationally known astronomy in the old Stone Age, this astronomy was factually adequate *as well as* emotionally satisfying, *it solved both physical and social problems* (one cannot say the same about modern astronomy) and it was tested in very simple and ingenious ways (stone observatories in England and in the South Pacific; astronomical schools in Polynesia – for a more detailed treatment and references concerning all these assertions cf. my *Einführung in die Naturphilosophie*). There was the domestication of animals, the invention of rotating agriculture, new types of plants were bred and kept pure by careful avoidance of cross fertilization, we have chemical inventions, we have a most amazing art that can compare with the best achievements of the present. True, there were no collective excursions to the moon, but single individuals, disregarding great dangers to their soul and their sanity, rose from sphere to sphere to sphere until they finally faced God himself in all His splendour while others changed into animals and back into humans again (cf. Chapter 16, footnotes 20 and 21). At all times man approached his surroundings with wide open senses and a fertile intelligence, at all times he made incredible discoveries, at all times we can learn from his ideas.

Modern science, on the other hand, is not at all as difficult and as perfect as scientific propaganda wants us to believe. A subject such as medicine, or physics, or biology appears difficult only because it is taught badly, because the standard instructions are full of redundant material, and because they start too late in life. During the war, when the American Army needed physicians within a very short time, it was suddenly possible to reduce medical instruction to half a year (the corresponding instruction manuals have disappeared long ago, however. Science may be simplified during the war. In peacetime the prestige of science demands greater complication.) And how often does it not happen that the proud and conceited judgement of an expert is put in its proper place by a layman! Numerous inventors built 'impossible' machines. Lawyers show again and again that an expert does not know what he is talking about. Scientists, especially physicians, frequently come to different results so that it is up to the relatives of the sick person (or the inhabitants of a certain area) to decide *by vote* about the procedure to be adopted. How often is science improved, and turned into new directions by non-scientific influences! It is up to us, it is up to the citizens of a free society to either accept the chauvinism of science without contradiction or to overcome it by the counterforce of public action. Public action was used against science by the Communists in China in the fifties, and it was again used, under very different circumstances, by some opponents of evolution in California in the seventies. Let us follow their example and let us free society from the strangling hold of an ideologically petrified science just as our ancestors freed *us* from the strangling hold of the One True Religion!

The way towards this aim is clear. A science that insists on possessing the only correct method and the only acceptable results is ideology and must be separated from the state, and especially from the process of education. One may teach it, but only to those who have decided to make this particular superstition their own. On the other hand, a science that has dropped such totalitarian pretensions is no longer independent and self-contained, and it can be taught in many different combinations (myth and modern cosmology might be one such combination). Of course, every business has the right to demand that its practitioners be prepared in a special way, and it may even demand acceptance of a certain ideology (I for one am against the thinning out of subjects so that they become more and more similar to each other; whoever does not like present-day Catholicism should leave it and become a Protestant, or an Atheist, instead of ruining it by such inane changes as mass in the vernacular). That is true of physics, just as it is true of religion, or of prostitution. But such special ideologies, such special skills have no room in the process of *general education* that prepares a citizen for his role in society. A mature citizen is not a man who has been *instructed* in a special ideology, such as Puritanism, or critical rationalism, and who now carries this ideology with him like a mental tumour, a mature citizen is a person who has learned how to make up his mind and who has then *decided* in favour of what he thinks suits him best. He is a person who has a certain mental toughness (he does not fall for the first ideological street singer he happens to meet) and who is therefore able *consciously to choose* the business that seems to be most attractive to him rather than being swallowed by it. To prepare himself for his choice he will study the major ideologies as *historical phenomena*, he will study science as a historical phenomenon and not as the one and only sensible way of approaching a problem. He will study it together with other fairy-tales such as the myths of 'primitive' societies so that he has the information needed for arriving at a free decision. An essential part of a general education of this kind is acquaintance with the most outstanding propagandists in all fields, so that the pupil can build up his resistance against all propaganda, including the propaganda called 'argument'. It is only *after* such a hardening procedure that he will be called upon to make up his mind on the issue rationalism-irrationalism, science-myth, science-religion, and so on. His decision in favour of science – assuming he chooses science – will then be much more 'rational' than any decision in favour of science is today. At any rate –

science and the schools will be just as carefully separated as religion and the schools are separated today. Scientists will of course participate in governmental decisions, for everyone participates in such decisions. But they will not be given overriding authority. It is the *vote* of *everyone concerned* that decides fundamental issues such as the teaching methods used, or the truth of basic beliefs such as the theory of evolution, or the quantum theory, and not the authority of big-shots hiding behind a non-existing methodology. There is no need to fear that such a way of arranging society will lead to undesirable results. Science itself uses the method of ballot, discussion, vote, though without a clear grasp of its mechanism, and in a heavily biased way. But the rationality of our beliefs will certainly be considerably increased.

Name Index

Abbé, E., 138n
Abraham, M., 39n
Achinstein, P., 252n
Aenesidemus, 153n
Aeschylus, 249
Agatharchos, 249
Akiba, Rabbi, 190
al-Farghani, 110
Alhazen, 132
Althusser, L., 147n
Ames, A., 130n
Anaximander, 57, 246
Anaximenes, 86n
Aquinas, St Thomas, 274n
Aristarchus, 56, 101–2, 307
Aristotle, 35n, 49, 58 and n, 85n, 86n, 95, 109, 112, 118, 148n, 149, 210, 213n, 224, 242n, 257n, 277, 283, 305
Armitage, A., 92n
Armstrong, D., 164
Ashmole, B., 230n, 231n
Augustine, St, 65n
Austin, J. L., 79n, 247n, 261n
Autolycus, 133
Ayer, A. J., 247n, 258n

Bacon, F., 73, 76, 157n, 223
Bacon, R., 118
Bakunin, M. A., 187n
Barrow, I., 60 and n, 137n
Baumker, C., 121n

Beazly, J. D., 230n, 231n
Becher, J. R., 241
Becker, R., 39n, 62n
Bellarmine, R. F. R., 192–3
Benedetti, G. B., 94
Benn, G., 219
Berellus, 108n
Berkeley, G., 60n, 137n
Berossos, 132n
Besso, M., 57n
Birkenmajer, A., 96n
Blumenberg, F., 148n, 195n
Bohm, D., 43n
Bohr, N., 24n, 56, 62n, 64n, 164, 183n, 207, 253, 271, 275, 276, 279
Boltzmann, L., 35n, 40n, 89, 188n
Bondi, H., 282n
Born, M., 57n, 202n, 219
Bousset, W., 190n
Brahe, T., 192
Brecht, B., 17
Broderick, J., 193n
Brodsky, S. J., 61n
Brouwer, L. E. J., 258
Brower, D., 56n
Bruno, G., 49n, 92n, 96n, 111
Bub, J., 43n
Buck, R., 283n
Buchdahl, G., 103n
Bultmann, R., 47
Bunge, M., 111
Burmeister, K. H., 111
Butterfield, H., 17n
Butts, R., 213

Cannon, W. H., 50n
Cantore, E., 125

Carnap, R., 112, 158, 185n, 279, 280, 281n, 284
Carioso, 107n
Cartailhac, E., 231n
Carlos, E. St., 104n, 105n
Caspar, M., 124n, 142n
Cassini, G. D., 127n
Cassirer, E., 299
Castaneda, C., 190
Cesi, Cardinal, 107n
Cesi, F., 107n
Chazy, J., 56n, 57n
Cherniss, H., 130n
Chiaramonti, 71
Chiu, H. Y., 56n
Choulant, L., 132n
Christina, Grand Duchess, 101n, 106n
Chwalina, A., 128n
Cicero, 154
Clagett, M., 92n, 102n
Clark, K., 226n
Clavius, 107n, 126n, 127n
Clemence, G., 56n
Colodny, R. G., 164
Cohn-Bendit, D., 21n
Comte, A., 100n, 259n
Cook, J. M., 64n
Copernicus, N., 12, 13, 49, 52, 56, 69, 74, 84–5, 88 and n–90, 92, 95–7, 99, 101–3, 106n, 110, 111, 113, 114, 133, 138n, 139, 141–2, 146, 151–4, 156, 158–60, 178n, 191, 193, 202, 209n, 210 and n, 304, 305
Cornford, M. F., 297
Crassi, H., 128n
Crew, H., 84n
Croizier, R. C., 50n
Crombie, A. C., 105n, 298n

Dardel, 297
da Vinci, L., 130n
d'Elia, P. M., 127n
Democritus, 100, 161, 210
de Salvio, A., 85n
de Santillana, C., 49n, 50n, 296n
Descartes, R., 69, 70n, 116, 137n
de Witt, B. S., 282n
de Witt, C., 282n
Dicke, R. H., 56n, 57
Diehl, C., 268n
Diels, H., 261n
Dingler, H., 173
Dini, P., 107n
Diogenes of Sinope, 78
Dirac, P. A. M., 30, 35, 64n, 254
Dodds, E. R., 134n, 243n, 244n, 246n, 262n
Donati, L., 105
Dorling, J., 155n
Drabkin, I. E., 94n
Drake, S., 55n, 94n, 97n, 103n, 106n
Drell, S. D., 61n
Dreyer, J. L. D., 132n
Duhem, P., 35n, 109, 110, 146n, 188n
Düring, I., 148n, 257n

Eccles, J. C., 305
Edwards, P., 36n
Ehrenfest, P., 56n, 62, 202
Ehrenhaft, F., 39 and n, 40n, 62, 202
Ehrismann, T., 130
Einstein, A., 18 and n, 35n, 40 and n, 41n, 43n, 56n, 57n, 58n, 118, 164, 178n, 201–2, 207 and n, 213n, 219, 272n, 275, 278, 298
Else, G., 268n
Empedokles, 261n
Erasmus, 100n

Euclid, 83n
Eudoxus, 133, 173
Evans-Pritchard, E. E., 250n, 251 and n, 269, 272n, 296n, 298n, 302
Exner, F. M., 41n

Faraday, M., 89
Feigl, H., 57n, 165–7, 280
Festugière, A. M. J., 190n
Feynman, R., 61n
Fontana, F., 126n, 130n
Frankfort, H., 296n, 297
Frazer, A. C., 60n
Freundlich, E. F., 57n
Fugger, G., 104n
Fürth, R., 40n

Galileo, 12, 13, 14, 26, 35, 36, 55 and n, 61n, 64, Ch. 6–12, 162, 194n, 207, 284, 305
Gentile, G., 92, 111
Geymonat, L., 103n, 105n, 106
Giedymin, J., 253–5, 272n, 280n
Giuducci, M., 138n
Gombrich, E., 132n, 226n, 235n
Gonzaga, Cardinal, 107n
Gottschaldt, K., 122n
Grazioso, P., 231n
Gregory, R. L., 122n, 123n, 130n, 227
Griaule, M., 296n
Grienberger, 126n
Groenewegen-Frankfort, H. A., 235n
Grosseteste, R., 112, 126n
Grünbaum, A., 59n
Gullstrand, A., 137n
Guthrie, W. K. C., 261n

Habermas, J., 175
Hamilton, E., 296n

Hammer, F., 104n, 142
Hampl, R., 233n
Hanson, N. R., 38n, 103n, 133, 166, 237n, 264n, 269n, 277n
Harnack, H., 200n
Harrison, J., 297
Hanfmann, G. M. S., 236n
Hawkins G., 49n
Heaviside, O., 40n
Hegel, G. W. F., 18n, 27n, 78n, 257n, 258
Heiberg, J. L., 133, 175
Heilbron, J. L., 62n
Heisenberg, W., 38n, 58 and n
Heitler, W., 61n
Hempel, C. G., 278n, 280, 281n, 284
Henry of Hesse, 110
Heraclitus, 246n, 261n, 267
Heracleides of Pontos, 173
Herder, J. G., 229n
Hermes, Trismegistos, 96n
Herodotus, 245n
Herschel, J., 128n
Herwarth, 142n
Herz, N., 205n
Hesiod, 246
Hesse, M., 48, 49n
Hoffmann, W. F., 56n, 278 and n
Hipparchus, 109
Holton, G., 57n
Homer, 238–48, 260n, 261n, 265n, 268n
Hooke, R., 38n
Hoppe, E., 105
Horky, 123
Horton, R., 93n, 296–8
Hörz, H., 58n
Huebner, K., 145n
Hume, D., 65, 175

Hume, E. H., 51n
Huyghens, C., 105n

Ibsen, H., 21n, 302
Infeld, L., 278 and n
Iranaeus, 200n

Jacobsen, T., 296n
Jammer, M., 56n, 86n
Jansen, Z., 108n
Jeremias, 298n
Jones, R. F., 154n
Julmann, M., 110

Kalippus, 133
Kant, I., 66, 70 and n, 73, 106n, 173
Kästner, A. G., 105n, 121n, 130n
Kaufmann, W., 56 and n, 202
Kenner, H., 249n
Kepler, J., 35, 59 and n, 92, 104, 105, 113, 115, 118, 123–4 and n, 126n, 127n, 130n, 135–8n, 202, 205n
Keynes, J. M., 49n
Kierkegaard, S., 26, 175
Kilpatrick, F. P., 122n
Kirk, G. S., 239n
Klaus, G., 96n
Kock, S., 166
Koertge, N., 33n
Koffka, K., 122n
Kohler, I., 130
Kopal, Z., 128n, 129n, 130n, 135n
Körner, S., 43n, 273n
Köstler, A., 125n, 200n
Koyré, A., 92n, 96n, 157n, 277n
Krafft, F., 109, 246n
Kranz, W., 261n

Krieg, M. B., 51n
Kropotkin, P. A., 21 and n, 302
Kuhn, T. S., 37 and n, 62n, 166, 198, 298n
Kühner, R., 240n
Kurz, G., 241, 242n
Kwok, D. W. Y., 50n

Lactantius, 86n
Laertius, D., 30
Lagalla, J. C., 107
Lakatos, I., 5, 6, 14, 48n, 56n, Ch. 8, 153n, 176n, 177n, Ch. 16, 210n, Appendix 4, 259n, 272n, 284
Lattimore, R., 242n
Laudan, L., 103n
Leibniz, G. W., 281n
Lenin, V. I., 13, 17 and n, 18 and n, 145, 147n
Leopold of Toscana, 70n
Leo X, 106n
Lerner, M., 158n
Leroc-Gourhan, A., 231n
Levi-Strauss, C., 50n, 302
Liceti, 105
Lindberg, D., 115, 125n, 134n
Loewith, K., 257n
Lorentz, H. A., 43n, 56n, 60, 62, 178n, 202
Lucretius, 132 and n
Lysenko, T. D., 51, 306

Mach, E., 146n, 188, 259n
McGuire, J. E., 49n
Machamer, P. K., 103n, Appendix 2
McMullin, E., 32n, 85n, 126n, 127n
Maestlin, M., 104n, 130n
Magini, 123, 126n
Maier, A., 96n
Malavasia, 107n

Malinowski, B., 297
Manitius, C., 49n
Mann, F., 51n
Mao Tse-Tung, 147n
Marcuse, H., 27n
Marshack, A., 49n, 50n
Marwick, M., 93n, 297
Marx, K., 13, 145, 146n
Marzke, 279n
Matz, F., 230n, 233n
Maurolycus, 115, 136n
Maxwell, J. C., 39n, 60, 207
Meiner, F., 169n, 213n
Meisenheim, D., 61n
Mersenne, P., 70n
Meyer, A. C., 147n
Meyer, H., 257n
Meyerson, E., 146n
Michelet, C. L., 78n
Michelson, A. A., 56, 57n, 202, 282
Mill, J. S., 14, 20 and n, 48, 53, 158, 171, 173, 188n, 224n
Miller, D. C., 56 and n
Monaldesco, P., 107n
Moritz, Prince, 108n
Morley, E. W., 56, 57n
Morley, H., 108n
Musgrave, A. E., 56n, 176n

Nader, S. F., 65n
Nagel, E., 280, 284
Nakayama, T., 51n
Nestroy, 181
Neurath, O., 168, 169n
Newton, I., 29, 35 and n, 36, 49 and n, 56, 59 and n, 60, 63, 89, 90, 201–2, 207, 213n, 272n, 277, 283, 298
Nilsson, M. P., 246n

Nin, A., 157n

Page, D. L., 238n, 240n
Pappworth, M. H., 188n
Paracelsus, 299, 305
Pardies, P., 38n
Parmenides, 58 and n, 78, 118, 261n, 267
Parry, A., 267n
Parry, M., 238n, 239n
Pecham, J., 115–16, 119, 125n, 127n
Perrin, J., 40, 41n
Persio, 107n
Pfuhl, E., 247n, 268n
Philo, 153
Philolaos, 68n, 183n, 304, 305
Piaget, J., 227
Piffari, 107n
Pirandello, L., 33n
Plato, 87, 172–3, 227n, 257n, 263n
Pliny, 86n
Plinius, 109
Plutarch, 30, 130n, 133, 134, 268n
Poincaré, H., 56n, 146n, 173, 178n, 202
Polanyi, M., 166
Polemarchus, 133
Polyak, S. L., 124n, 138n
Popper, K. R., 14, 26, 35n, 41n, 48, 57n, 93, 111, 112, 169n, 171, 213n, 225n, 232n, 237n, 258, 273n, 274, 283n, 284, 298
Porphyry, 65n
Post, H. R., 60n, 273n
Pribram, K., 265
Price, D. de S., 102n, 110
Proklos, 109
Prout, W., 183n
Ptolemy, 49 and n, 84n, 86n, 92n, 103n, 109, 114, 210, 305
Pythagoras, 49

Raabe, P., 241
Radner, M., 165n
Radnitzky, G., 61n
Ratliff, F., 138n
Rattansi, P. M., 49n
Regiomontanus, 110
Rheticus, G. J., 110, 111
Richter, C. R., 50n
Richter, J. P., 130n
Riedel, J., 257n
Righini, G., 117
Rock, I., 130n
Ronchi, V., 59n, 105n, 122n, 123n, 125n, 129n, 130n, 134n, 136 and n, 138n
Rosen, E., 49n, 88n, 104n
Rosen, S., 70n
Rosenfeld, L., 24n, 43n, 61n, 203n, 279n
Rosenham, D. L., 188n
Rosental, S., 24n
Rothmann, J. P., 118
Rubin, E., 258n
Rutherford, E., 202

Salmon, W., 58n
Sambursky, S., 133
Schachermayer, F., 245n
Schäfer, H., 231n, 232n, 234n, 235n, 237n, 247n
Schulz, W., 130n
Schumacher, C., 103n
Schumann, F., 51n
Schwarzschild, C., 63
Schweitzer, A., 194
Scott, 96n
Seelig, K., 57n
Sen, D. H., 60n, 278n

Seznec, I., 195n
Shankland, R. S., 57n
Shao, C., 50n
Shapere, D., 270n, 277
Shaw, G. B., 21n
Sherif, M., 22n
Simplicius, 118, 133
Simon, G. M., 219
Simonides, 268n
Skinner, B. F., 48n
Smart, J. J. C., 164
Smith, K. W., 130n
Smith, W. M., 130n
Snell, B., 134n, 242n, 245n, 247n, 260n, 261n, 262n, 265n
Solon, 268n
Sonnefeld, A., 126n
Steneck, N. H., 134n
Stratton, G. M., 130, 228n
Strawson, P. F., 115
Strindberg, A., 21n
Stroud, B., 280n
Stuewer, R., 49n
Summers, A. J.-M. A. M., 274n
Svedberg, T., 40
Synge, J., 282n
Szentgyorgi, 188n

Tarde, J., 104, 105n
Taylor, H., 48n
Terrentius, 107n
Thales, 86n, 246n
Thutmosis, 237
Tillich, P., 47
Tillyard, E. M. W., 86n
Tolansky, S., 125n
Toscana, Duke of, 124

Toulmin, S., 132
Tranekjaer-Rasmussen, E., 258n
Trotsky, L., 147n
Truesdell, C., 36n

Watkins, J. W. N., 157n
Webster, T. B. L., 230n, 232n, 233n, 238n, 239n, 240n
Wheeler, J. A., 279n
White, J., 268n
Whorff, B. L., 223–4, 237, 270n, Appendix 5
Wieland, W., 148n, 213n
Wigner, P. E., 64n, Appendix 3
Winokur, S., 165n
Witelo, 83n, 127n
Wittgenstein, L., 133
Wohlwill, E., 124n
Wolf, R., 105n, 126n, 129n, 135n
Wolff, R. P., 21n

van der Waerden, B. L., 208n
Veith, I., 51n
Velikovsky, E., 40n, 298n
Vernon, M. D., 122n, 130n
Vives, 100n, 121n
von Dechend, H., 50n
von Dyck, W., 124n, 142n
von Fritz, K., 68n
von Helmholtz, H., 137n
von Hoddis, J., 241
von Kleist, B. H. W., 257n
von Köln, E., 126n
von Nettesheim, A., 108n
von Neumann, J., 30, 84n, 253, 260n
von Rohr, M., 137n
von Smoluchowski, M., 41n
von Wilamowitz-Moellendorf, U., 244n, 245n

von Soden, W., 263n

Xenophanes, 86n, 210, 262, 264n

Yates, F., 49n

Zahar, E. G., 43 and n, 178n, 210n
Zeno, 58 and n, 78, 242
Zilboorg, G., 100n
Zinner, E., 105n, 109, 110, 126n, 129n, 130n

Subject Index

t = *term explained*

action, 23–6, 45, 168, 172, 182, 186–7, 196–7, 211, 221–2, 248, 256, 260, 262, 265, 271, 307; see under ideas; standards

acupuncture, 50, 305; see also Chinese communism and medicine

ad hoc hypotheses: and critical rationalism, 172, 177–9; and incommensurability, 278; and myth, 93n, 297; presence in modern science, 41n, 42n, 59, 63–4; progressive role in science, 12, 23, Ch. 8, 99, 129n, 143, 154–6, 167, 178

aim of science, 30, 195, 208, 296, 299

alienation, 261

anamnesis, 12, 73, 81, 87–9, 143

anarchism: epistemological, 2, 3, 11, 14, 17, 21 and n, 32–3, 165, 171, 175, 180, 181, 187n, 214, 221; political, 20, 21 and n, 187–9; religious, 187, 189; see also dadaism

anthropology, 49, 93n, 205, 280; the anthropological method and incommensurability, 269, 271, 272 and n; of science and cosmology, 184n, 249–60, 266n, see also field study

anything goes, principle of, 11, 28, 33, 186, 197, 215, 296

appearances, 72, 134, 258n; in ancient Greek cosmology, 260–9; reality or fallacy of, 71, 74, see also natural interpretations; vs. reality, 66

argument: and cosmology, 232, 256; vs. emotions, 295–6; and epistemological anarchism, 159, 189, 191–2, 194–5; as a hindrance to progress, 24, 195; and incommensurability, 80, 171–2, 225, 272n; the limited value of, 25, 81, 153–4, 210; and logicians, 184n, 260; as a method of indoctrination, 24, 200, 308; and the methodology of scientific research programmes, 197–8, 200–1; from observation and natural interpretations, 74, 79, 88, 99; and scientific chauvinism, 220

Aristotelianism, 209t; basic value judgments, 207–8; vs. Copernicanism, Ch. 6–12, 69, 88, 160, 209 and n, 210, 211; dynamics and theory of motion, 95 and n, 96, 99, 150, 161, 224; empiricism, scientific method, and theory of knowledge and perception, 35n, 58n, 89, 112, 115, 121 and

n, 122 and n, 148 and n–50, 213n; form of life, 163, 192; philosophical system and cosmology, 148–50, 154, 207, 283; science, 61n, 156, 181, 184, 205, 277; theory of space, 225
Aristotle's theories, astronomy, 109; theory of the continuum, 88n
art, 52, 135 and n, 146n, 147n, 247–8, 250, 268, 271, 307; archaic style, 230–45; and science, 52, 166, 167
astrology, 100n, 205–6, 208n, 303
astronomy, 49, 100n, 109–11, 149, 193–5, 210, 211, 225, 301, 304–6; ancient Greek, 49n, 208n, 263n; Babylonian and Egyptian, 208n, 263n; medieval, 208n, 209n, 213n; palaeolithic and stone age, 49n, 50n; Ptolemaic, 88n, 103n, 109–11, 113–14, 159, 210n; and science, 52, 166, 167; see also Copernicanism
atomism, 52, 56, 86n, 183, 186n, 212
authority, 33, 181, 188, 200, 216, 299, 304, 309
auxiliary theories and sciences, 44, 66–7, 99, 126n, 151–4, 157, 278; see also secondary elaborations
axiomatics, 255

background knowledge, 66–7, 213n
basic statements, 65, 172, 201, 208n, 225n
basic value judgments, 201t–10
belief(s), 19, 26, 56, 71, 211; basic, 297, 309
big shots, 303, 309; see also top dogs; Nobel prize winners
biology, 100, 147, 283, 307
botany, 61n, 306
Brownian motion, 39–41n, 43n
business, see under science, modern science like business

chaos, 21, 179, 181, 196, 262, 268
Chinese communism and medicine, 50–2, 220, 305, 307; see also science political interference in
Chinese lady's foot, 20
Christology, 195
church, 15, 52, 86n, 106n, 200n, 216, 295, 299, 301; see also state
clashes, 32nt; see also suspends
classical mechanics, 146, 271, 276, 278, 282, 284

classifications, 274, 276, Appendix 5, 297, 298, 306; covert, 224t and n, 225, 270, Ch. 17; see also incommensurability; instrumentalism
common sense: and Copernicanism, 83, 86 and n, 89, 99, 160; and incommensurability, 256, 266, 271, 281; and materialism, 164; conservativism and the methodology of scientific research programmes, 199 and n–204, 207n; science and myth, 297; see also natural interpretations
complementarity, principle of, 42–4
concepts, 67–8, 76, 85, 88, 279n, 280, 285; criticism of, 67–8; see also counterinduction; natural interpretations
conceptual change, 70, 164, 225, 232, 256–7, 265–6, 268–9, 275–7
conceptual continuity, 277 and n, 283–4
conceptual discovery, 269
conceptual progress, 283
conceptual totalitarianism, 262
confirmation, 29, 38, 44, 65, 91, 145, 282
conjectures and refutations, 143, 174, 259, 270
consistency, 114, 184n, 260; condition, Ch. 3; vs. counterinduction, Ch. 2
content, see empirical content
context of discovery vs. context of justification, 165–7
continuity, 276–7n, 283–4
conventionalism, 207, 280n, 287
Copernicanism, 12, 13, 14, 23, 26, 30, 49, 52, 55, 66–7, Ch. 6–12, 163, 178n, 186n, 191–5, 202, 207, 209n–11, 301, 304–5
correspondence, principle of, 64, 103n, 276
correspondence rules, 75, 255
corroboration, see confirmation
cosmology, 26, 67, 71, 92, 111, 149–52, 206, 208n, Ch. 17, 302, 304, 308; alternatives and counterinduction, 31, 47n; anthropological study of, 249–52; changes of, Ch. 17; covert classifications and incommensurability, 224–5, Ch. 17, 284; and elimination of degenerating research programmes, 185n–6n; and language, 223–4; and methodology, 185n, 186n, 195, 206–7, 295–6, 304
counterinduction, Ch. 2–3, 47, 66–8, 77–8, 103
courtesans, 282; see also prostitution
cranks and madmen, 47, 68

critical rationalism, 156 and n, Ch. 15, 182, 308
criticism, 31–2, 42, 68, 73, 112, 152, 160, 165–6, 171–3, 185, 206, 229, 272n
crucial experiments, 40, 43, 282, 284

dadaism, 21n, 33n, 189 and n, 191; see also anarchism
demarcation between science and non-science, 48, 201, 212
democracy, 301–3, 307, 309
demonic possession, 44, 100n, 276; see also voodoo; witchcraft
Descartes' philosophy, 86, 213n, 275
development of culture, 30, 180
development of the individual, 11, 15, 20, 35, 45, 52, 175, 180, 187, 193, 217–18, 295; see also education; freedom; pluralism
dialectic: Hegelian, 27; Popperian, 26, 171–6, 274
discovery, 27, 43, 114, 146, 154, 165–7, 172, 174, 184, 267–9, 275, 296, 307; context of, 165–7; incommensurable, 269; of natural interpretations by counterinduction, 75–8
dogmatism, 42, 70n, 100, 111, 169, 176n, 181, 257–9, 295, 298

education, 19, 20, 24, 39, 45, 52, 187, 217, 237n, 256–7, 299–301, 303, 308
Einstein's methodology, 57n, 58n, 213n
Einstein's theories, see relativity
emotions, 154, 220, 265 and n, 296, 306
empirical content of theories, 172t; and the autonomy principle, 38–9; comparisons of, 177, 212, 214, Ch. 17, 224, 232, 270, 278, 284; decreased, 23–4, 113, 153, 156 and n, 172, 176–7, 186, 268, 278; epistemological illusion, 177–9, 195, see also epistemological illusion; and ideal languages, 280n; increased, 30, 41, 47–8n, 93–4, 159, 172, 176–8, 195, 211–12, 296; lack of, 183 and n
empirical support, 37, 42, 43, 156, 260, 282; see also confirmation; corroboration
empiricism, 29t, 46 and n, 70n, 89, 159, 228, 279, 295; Aristotelian empiricism, see Aristotelianism; contemporary, 37, Ch. 14; critical, 172–80; demand for increased empirical content, 41, 212, see also empirical content; logical, 179t, 282, see also logical positivism; see also experience
epistemology, 17, 18, 23, 89–90, 149, 185n, 204, 213 and n, 229n, 259; epistemological anarchism, see anarchism; epistemological dadaism, see

dadaism; epistemological illusion, 177–9, 186n, 195 and n, 211–28n, see also empirical content; epistemological opportunism, see opportunism; epistemological prejudices, 66; epistemological prescriptions, 20, see also methodology (b)
equipartition principle, 62 and n
error, 155, 157, 179, 206
essentialism, 261n–71
evaluation of theories, see methodology (a)
evidence, 11, 26, 31, 44, 65, 67, 68n, 93, 112–13, 139, 141–2, 153, 156–8, 176, 228, 250; manufactured, 26, 44, 99, 211
evolution: of a theory, 103; theory of, 30, 188, 307
examinations, 217–18
existentialism, 218
expectations, 172–4; see under dialectic, Popperian
experience, 29, 46n, 65, 87–9, 168–9, 261, 286; Aristotle, 149; changes in experience to fit theory, 87–9, 91n–2 and n, 99, 100, 106, 121; metaphysical, speculative, 92, 100
experiment, 42, 55, 58, 64, 67, 91, 107, 172, 206
experimental results, 29, 31, 37, 66, 183 and n
experts, 19, 30, 58, 182, 217, 305
explanation, 43, 44, 177n, 178, 184, 246, 269, 283–4, 297, 303; see also *ad hoc* hypotheses

facts, 29, 46, 55, 64–5, 77, 100, 102, 112, 163, 192, 269, 272n, 284, 286–7, 298–9; and autonomy principle, 38; clashes between theory and facts, 12, 31, Ch. 5, 113–14; collection, discovery, and suppression of facts, 39–44, 154, 160, 174, 176–8, 257–8; and incommensurability, 274–6; novel, 37, 43n, 102, 209n, 210, 287; theoretical nature, 19, 31, 39, Ch. 5; see also counterinduction; evidence; experience; experiment; natural interpretations
fairy-tales, 30, 32, 52, 156, 209, 210, 271, 273, 300, 302–4, 306, 308
faith, 103, 154, 192, 305
falsificationism: ahistorical, 145 and n; vs. anthropological field work, 260; and cosmological assumptions, 295; vs. counterinduction, 32, see also counterinduction; and critical rationalism, 173, 179; and discovery, 43n; and Einstein's methodology, 57n, 213n; eliminated, 66; eliminates

science, 65, 176, 303; and historical method of evaluating methodologies, 201, 205–7; and history of science, 182; and irrefutability of theories, 114, 303; vs. methodology of scientific research programmes, 186, 198; naive, 57n, 176, 198, 205–6, 297, 303; and principle of proliferation, 48n; sophisticated, 48n
field study: and incommensurability, 272–3, 282; of science, 260, 298; see under anthropology
formalism, 37, 64n, 169; in aesthetics, 235–6, 238, see under art, archaic style; formal systems, 158, 183 and n, 252–7, see also logic; formalists, 258
forms of life and thought, 76, 163–4, 189, 193, 216, 219, 252, 258, 295, 299; see also field study; anthropology
frameworks, 67, Ch. 17, 233, 264, 269, 271, 273n, 277; see also cosmology; concepts; incommensurability
freedom: and anarchism, 187; and Copernicanism, 154; and critical rationalism, 175; from science and systems of thought, 175, 268, 285, 299, 307; of association, 187, 216; of choice in science, 303, 308; of speech and debate for science, 21; of will, 175; and reason, 179–80; and scientific chauvinism, 306–7

Galileo: and Copernican revolution, 55, Ch. 6–12; dynamics and mechanics, 35–6, Ch. 7, 94–7, 99–100, 103, 143, 153n, 160–1; method, 14, 81, 87, 97–9, 102–3, 112–13, 159–61, 163; and the moon, 117, 127–135; optics and the telescope, 13, 99, 103–8, 114–19, 122n, 129–31, 135–6, 138 and n–9, 142–3, 160; relativity, 74–5, 78–9 and n, 81–92, 95, 97, 160, 163; and tower argument, Ch, 6–7, 70–1, 75, 84, 90, 97, 146
Genesis, 30
governmental decisions, 301, 309; see also democracy
grammar, 86n, 163, 223, 232, 238, 270, 273, 278, 286
growth of knowledge and science, 23, Ch. 8, 167, 174, 296; and the methodology of scientific research programmes, 184 and n; without argument, 24

happiness, 21n, 48n, 175, 179, 180
hermeticism, 49 and n, 190n
hidden variables, 147, 209n

history, 17 and n, 18 and n; backward movements in history, 153 and n, 268; economic, social, political, 17–18, 146n–7n, 206n; evaluating the history of science, 184, 186, 196, 210n; historical background of science, 66–7, 145, 154, 187n, 210, 295; historical method of evaluating methodology, see under methodology (a); historical refutations of methodology, 23, 27–8, 67, 143, 159, 176, 179; historical research programmes, 186n, 214; history of art, 231; history of science, 19, 27, 30, 49n, 161, 296; history of science in education, 301, 308; history of science and incommensurability, 171, 253 and n, 258–9, 269, 271, 287; history of science and philosophy of science, see under philosophy of science; internal/external, 166, 197, 210–12; see also rational reconstructions
hole or Swiss cheese theory of representation, 266
Homo Oxoniensis, 45n, 229n
humanism, 218
humanitarianism, 17, 20, 46, 48, 52, 188, 191
humour, see under progress, aids to
hydrodynamics, 278
hypotactic systems, 236

ideas, 11, 114, 175, 273, 286, 300, 306–7; and action, 25–7; anthropology and key ideas, 250
ideology, 15, 26, 43–4, 55, 67, 72, 77, 164, 176–7, 187, 189, 193, 200, 204–205, 207, 208 and n, 210, 211, 214, 232, 245, 250, 251, 272, 274, 278, 295, 299–308; see also fairy-tales; scientific method
ignorance, 44, 48, 52, 112, 211, 260, 297
Iliad, 47n, 239n–248, 261, 267
imagination, 14, 45, 72, 91, 217, 232, 268
impetus theory, 79, 95–7, 225, 277 and n
incommensurability, 30, 114, 142n, 158, 165, 171, 178, 214, Ch. 17, 224, 269t
inconsistency, 29, 35–6, 55, 183 and n, 186
incubi, 126n, 274
inductive logic, 183n–4n
inductivism, 103, 145, 173, 181, 186, 201, 207, 265
instrumentalism, 111, 193–4, 223, 250, 274, 279, 283
instruments, 26, 78, 105–7, 148, 151, 159, 187n, 210, 214, 217, 223; see also telescope

intellectual pollution, 217–20
intersubjectivity, 126n, 280n
intuition, 19, 169
irrationalism, 13, 25, 32, 154–5, 165, 172, 182, 195–9, 205, 208n, 209, 214, 219, 270, 308

justification, context of, 155, 165–7

Kepler: astrology, 205n, celestial essence, 118; laws and Newton's theory, 35, 202; and moon, 127n, 130n, 135n; optics and the telescope, 59 and n, 104 and n, 105 and n, 113, 155, 136 and n, 138n, 141; polyopia, 124n, 127n, 128n
kinetic theory, 39, 40, 41n, 43n, 147
knowledge, 20, 30, 46, 52, 77, 93, 145, 149, 154, 157, 168, 174, 179, 182, 188, 190, 196, 206, 208n, 211–14, 219, 229, 245n–8, 259–64, 298, 302–3, 306; see also lists as knowledge

language, 27, 66–7, 72–5, 79, 87, 151, 158, 223–4, 255, 262, 264, 271, 272n, 280 and n, 282, Appendix 5
law and order, 11, 17, 21, 27, 93, 165, 171, 181, 197, 200n
learning, 24–6, 52, 168, 176, 217, 234, 272, 280, 281
libertines, 219
liberty, 20, 26; see also freedom
linguistic philosophy, 154, 287
linguistic relativity principle, 286t
lists as knowledge, 263 and n, 270
logic, 19, 27, 48, 146, 158, 165, 182–4n, 223, 224n, 229, 232, 250, 252–60, 269, 280, 287, 303
logical transference, Popper's principle of, 258
logical positivism, 48n, 112, 179, 280 and n, 282
Lysenko affair, 51, 216

mafia, Lakatos's, 210
magic, 108n, 181, 205, 298, 299
Marxism, 146, 148
materialism, 163–4, 277

mathematics, 64, 66, 115, 230n, 254, 257, 303; ancient Greek and Babylonian, 263n
meaning, 229, 252 and n, 255, 273n, 279, 280; see also meaning
measurement, 40, 55, 56n, 63, 179
medicine, 50–2, 65, 100, 205, 284, 304–7
mental sets, 226, 264, 268
metaphysics, 19, 32, 52, 58, 88, 92, 100, 153, 160, 180, 282, 285
meteorology, 67, 151
methodology:
(a) *as standards of appraisal*: anarchistic standards, 187–96, see also anarchism; confirmation and corroboration, 65; and cosmological assumptions, 185n, 186n; and formalism, see formalism; intuition, 169; Lakatos's historical method of evaluating standards, 183, 201–9, 213; Lakatos's standards, 14, 15, 181, 184–7, 196–209n, 212–14, 223; and research funding, 216; unit of appraisal, 39, 183–4
(b) *as rules of scientific practice*: and anarchistic practice, 187–201, see also anarchism; anything goes; appraisal of, 205–6, 213, 295–6, and cosmological assumptions, 195, 206–7, 295–6, 304; of counterinduction, see counterinduction; and counterrules, 29–32; of critical rationalism, Ch. 15; democractic rules, see democracy; for elimination of theories and research programmes, 167–8, 182–3, 185 and n–186; of empiricism, see empiricism; enforcement, 198; of falsificationism, see falsificationism; of inductivism, see inductivism; and history of science, 17–19, 165–7, 183n–184n, 211; their limitations, 32; of logical empiricism, 179; of logicians, 260; and the methodology of scientific research programmes, 185, 187n; and opportunism, see opportunism; and politics, 18 and n, see also politics; principle of proliferation, see proliferation; and reason, 181–2; vs. scientific practice, 19, 23, 65–7, 143, 159, 167, 179–80, 182–3; and scientific problem solving, 302; and theory of rationality, see under rationalism; of the unique scientific method, see under scientific method; their violation and scientific progress, 23–4; unrealistic, pernicious, and detrimental to science, 295–6
methodology of scientific research programmes, see research programmes
283

Michelson-Morley experiment, 282
mind/body problem, 80, Ch. 13, 190, 262, 277
mob psychology, 199, 211
model theory, 254, 255, 297
money, 187n, 196–7, 216
morality, 24, 105–6, 180, 187, 245–6, 262n
myth, 30, 44–50 and n, 67–8, 171, 181, 218, 254, 261n, 271, 282; compared with science, 14–15, 44–5, 49, and n–50n, 52, 180, 184, 196, 208n, 223 245, 295–302, 308

natural interpretations, 69t, Ch. 6–7, 99, 103, 146, 151, 163, 164, 278
Newton's method and theories, 59, 201–2, 213n, 225, 228, 271n–2n, 277,
nonsense, 256–7, 270; see also meaning
novel facts, see under facts
Nuer, 251

objectivity, 19, 28, 46, 53, 67, 123, 152, 181, 190, 196, 302–3; see also intersubjectivity
observation: and *anamnesis*, 73; argument from observation, 74, 99; and counterinduction, 30–2, 55, 68; and Galileo, 91, 102, 123n, see under Galileo, and the moon, optics and the telescope; historical nature of, 145; and incommensurability, 230, 256, 265, 269, 275, 278–9, Appendix 5; intersubjective, 126n–7n; language, 66–7, 72, 79, 80, 81, 87, 255, 274, 279–82; and natural interpretations, Ch. 6–7, see also natural interpretations; observational laws, 35n; reality or fallacy of, 71–2; role in science, 26, 43n, 145–7, 157, 172–3, 176, 193–5, 206, 259; statements, 38, 67, 72, 76, 78, 278; production of observation statements, 74; sensory core of observation statements, 76, 100, 163; telescopic, see telescope; terms and theory, 45, 66, 68, 75, 165, 168–9 and n, 278–83; terrestrial vs. celestial, 117–19, 123 and n; theories of, 31, 148–52; to be ignored, 152–3; and witchcraft, 44; see also experiment, instruments, measurement, natural interpretations, perception, physiology, sense impressions
Odyssey, 240
ontology, 79, 176, 236, 244n, 249, 275, 281
opportunism, 175, 187; Einstein's 58n, 213n; epistemological, 18, 70n, 179

paratactic aggregates, 233t, 240 and n, 260, 263
passion, 26, 179, 191
perception, 31–2, 66–7, 72–6, 115–19, Ch. 10, 148–9, 153, 168, 187, 190, 207, 223, 225–33, 237–8, 243n, 244n, 249, 262–3, 265–7, 269 and n, 271–272, 274; see also natural interpretations; observation; perspective; physiology, sense impressions
perspective, 230n, 231n, 247 and n, 249, 262–3, 266, 268
phenomenology, 45, 258 and n
philosophy of science: and anarchism, 17; and anthropological method, 249–60; *a priori*, 203n; bastard subject, 301; and critical rationalism, 172, 229 and n; and history of science, 23, 47–48 and n, 61n, 145n–6n, 159, 165–7, 184n; humanitarianism and scientific education, 20; and paradigmatic cases, 40n; and politics, 18n; and rational reconstruction, 155n
physical objects, 67, 227–8, 263–4, 270, 275
physicalism, 212
physics, 39n, 40n, 61n, 65, 100n, 149, 152, 159, 193, 209–10, 225, 259, 280n, 301, 307–8; classical, 63, 66, 89, 152, 225 and n, 272n, 275–6n, 282–4; modern, 61 and n, 63, 206
physiology, 21–2, 50, 66, 100, 115–16, 124n, 126n, 133, 137n, 149, 151, 160, 187, 227, 231–2, 275; see also incommensurability; natural interpretations
pluralism, 14, 30, 47, 51–2, 154, 171, 203, 297, 298; see also proliferation
politics, 18, 25, 47, 51–2, 100n, 147n, 175, 187–9, 216, 257, 300–1, 303
Popperianism, 48, 57, 93, 111, 171–80, 207, 212–13n, 229n, 274, 276, 298
positivism, 100n, 171, 175, 276; see also logical positivism
prediction: novel, 97, 174, 303; numerically inaccurate, 55, 56 and n, 57 and n; and the task of the scientist, 30
prejudice, 31, 44, 47, 66, 73, 106, 154, 179, 209, 260, 300, 302
presocratics, 78, 132, 268n
presuppositions, 31, 145; *a priori*, 73
primitive thought, 64n–5n, 273
progress: historical, 146n–7n; scientific, see scientific progress
proliferation, principle of, 33, 48n, 51–2; see also pluralism
proof, 166

propaganda, 25, 81, 89, 99, 100, 106, 111, 143, 154, 157, 193, 200, 204, 209, 220, 303, 307, 308
prostitution, 24, 217, 308
Protestantism, 46, 48
psychoanalysis, 148, 244n
psychology, 75, 81, 100, 116–19, 125n, 133, 149, 165–6, 183n, 184n, 196, 206n, 209n, 211, 231, 258, 259 and n, 264, 275
Ptolemaic astronomy, see under astronomy
Puritanism, 21n, 25, 46, 48, 219

quantum mechanics, 23, 42, 61, 63–5, 156n, 178n, 179, 202, 224, 252–3, 260, 271, 275, 276n, 277n, 309

radical translation, see translation
rational reconstructions, 48n, 155n, 166, 201, 203, 207–8
rationalism, 25, 32, 48n, 51n, 70n, 81, 149, 165, Ch. 15, 181–2, 184–5, 190–192, 196–200, 205, 208–9n, 214–15, 217, 264, 272n, 285, 296, 308–9; theory of rationality, 25, 27, 165, 171, 179, 186, 196–201, 212, 214, Appendix 4, 302, see also methodology (b)
realism, 75, 86–7, 148, 151, 231, 237n, 249, 250, 274, 276, 279, 282
reality, see under appearances
reason, 20, 26, 33, 56, 73–92, 145, 154–6, 171, 179–82, 186–7, 190–1, 218, 261n, 295, 305
reconstructions, 253; anthropological, 249–55; logical, 250, 255, 264, 280n; rational, see rational reconstructions
reduction, 41n, 284
relativity: Galilean, see under Galileo; general, 56–8, 178n, 271, 278n, 279 and n; special, 56-8, 62–3, 178n, 202, 228, 251, 275–6, 278, 282; theory of, 65, 156n, 179, 224, 238, 259n, 281
religion, 19, 24, 68, 181, 196, 208n, 216–17, 245–6, 295, 298, 301, 307, 308
replies to critics, 33n, 48 and n–52, Appendix 2, 155n, Appendix 3, Appendix 6, 269n, 270n; see also wicked remarks
research, 13, 19, 26, 61, 89, 93, 99, 113, 125n, 129, 134, 166, 173, 178 and n, 187, 188n, 212, 219, 229, 245n, 249, 260, 298; anthropological research into science, 249–60, 270
research programmes, methodology of, 14, 178n, Ch. 16, 210n

rhetoric, 1–309, 32, 123n, 187, 204, 257
rules: and dialectic, 27; linguistic, 251, 256; socially restrictive, 268; see also anarchism; law and order; methodology (b)

scepticism, 73, 153n, 189, 297, 298
schools, see education
school philosophers, 175, 211
science: chauvinism of, see scientific chauvinism; democratisation of, 301–303, 307, 309; excellence of, 64, 298, 304; institutional, 24, 52, 174, 182, 187, 196–7, 199, 215–17, Ch. 18; modern science like business, 61n, 188, 197, 215–17; political interference in, 47, 51, 216, 305–7; and the state, see state
science fiction, 273
scientific change, 37, 183, 212, 272n, 278, 284, 295
scientific chauvinism, 12, 50, 51, 219–20, 306–7
scientific method, 49n, 125n, 155, 258, 279; and anarchists, 20–1, 302; and history of science, 23, 27, 64; as part of a theory of man, 175 and n; the unique scientific method as an ideological fairy-tale of scientific chauvinism, 220, 300–9; see also methodology: (b)
scientific practice, see under methodology: (b)
scientific progress: aids, methods, and sources of, 12, 13, 23, 27, 29, 37, 48–9, 50–2, 97–8, 100, 106n, 113, 145n, 152–8, 161, 174, 192, 195 and n, 210, 283, 302, 304–6; and anarchism, 11, 17, 27, 180; criteria and definitions, 27, 55, 156, 184, 276; hindrances and obstacles, 14, 24, 37, 75, 155–6, 165–6, 169, 176, 179, 182–3, 195, 260, 295, 303, 306; preconditions, 156, 167, 179, 184n, 195, 214, 255, 260, 268
secondary elaborations, 93n, 297t, 298
sensations and sense impression, 31, 66, 71–4, 76, 89, 99, 133, 149, 152, 168, 223n, 268; see also natural interpretations
senses, 31, 66, 71, 73–5, 78, 80, 90, 101, 103, 133, 149, 246 ,261n, 305; see also natural interpretations
simplicity, 24n, 113
simultaneity, 225
social peace, 191
sociology, 50, 165–7, 183n, 185n, 186n, 187n, 196, 206n
sophistry, 30, 69

speculation, 92, 100, 160–1, 297
spontaneity, 187
standards: of criticism, 32; and general education, 217–18; and liberty, 20; of rational action, Appendix 4; of rationality, see under rationalism, theory of rationality; see also methodology: (a)
state, 15, 21n, 52, 187, 216, 295, 299–302, 306, 308
status quo, 12, 25, 45, 47, 68, 152–3, 189, 194, 210
suspends, 32nt; see also clashes

taboo reactions, 297–8
technology, 299–300, 304
telescope, 13, 26, 67, 99, 103–8, 111, 114–19, Ch. 10–11, 158, 160, 194, 211
tests, 27, 38–44, 66, 78–9, 98, 106–7, 121, 129, 137, 143, 152, 159, 165–6, 230, 252, 283, 296, 300
textbooks, 37
theatre, 249, 268n
thermodynamics, 35, 39–41 and n
theology, 44, 46, 49, 100, 180, 200n
theoretical change, 176–9, 198
theoretical entities, 247n, 261, 265
theoretical support, 89
theoretical terms, 168–9, 279
theory, 98, 100, 146; and fact, 26, 55, 58, 59, 64, 143, 147, 178
third world, Popper's, 46, 159, 212, 285
top dogs, 40n, 65; see also big shots; Nobel prize winners
tower argument, Ch. 6–7, 97, 146; see under Galileo
translation, 270–83; radical, 287
trial and error, 105–6
truth, 27, 33, 43, 45, 81, 106, 169, 171, 179, 180, 186n, 189, 230, 261, 301–2

understanding, 26, 88, 250–2, 255, 306
uneven development, law of, 13, 145, 146 and n, 147 and n, 194, 206n
uniformity, 11, 19, 22n, 35, 45–6, 215
universities, 37, 219

verisimilitude, 284
voodoo, 48, 49–52, 283, 306

wicked remarks, 27, 40, 45n, 48, 114, 115, 119, 154, 169n, 171, 174–5, 181–2, 205, 208n, 210, 225n, 229 and n, 260, 264, 277n, 280n, 298–301
witchcraft, 44, 71, 93, 99, 100, 298

zoology, 306

Name Index and Subject Index compiled by Alex Bellamy